S. S. Rakhimkhodjaev
D. N. Kadyrova

TISSUE ENGINEERING THEORY

S. S. Rakhimkhodjaev
D. N. Kadyrova

TISSUE ENGINEERING THEORY

Complex weaves

ScienciaScripts

Cover image: www.ingimage.com

This book is a translation from the original published under ISBN 978-620-7-48661-8.

Publisher:
Sciencia Scripts
is a trademark of
Dodo Books Indian Ocean Ltd. and OmniScriptum S.R.L publishing group

120 High Road, East Finchley, London, N2 9ED, United Kingdom
Str. Armeneasca 28/1, office 1, Chisinau MD-2012, Republic of Moldova, Europe
Managing Directors: Ieva Konstantinova, Victoria Ursu
info@omniscriptum.com

Printed at: see last page
ISBN: 978-620-8-53127-0

Contents

OUTLINE.

The paper shows the theory of structure and peculiarities of dressing the production of fabrics of complex weaves. Definitions, purposes, advantages, disadvantages, principles, types of complex weaves are given. Methods of construction of complex weaves are given. The principles of construction of these fabrics are substantiated. Examples of construction and analysis of fabrics of complex weaves are given. Special attention is paid to the construction of the complete filling pattern of the weave on the basis of the fabric section, which simplifies the construction of complex weaves. It is intended for researchers, technologists, designers, constructors, dessinators, masters and bachelors, workers of textile industry dealing with complex weaves.

INTRODUCTION

Weaving as an art and craft has deep roots. In distant times, man created various objects for the convenience of existence - clothes, shoes, bedding, baskets, nets, etc. Obtaining these objects (products) was carried out by weaving strips of animal skin, grass, reeds, flax, bushes and trees. Such creativity of ancestors created weaving - one of the forms of weaving and weaving devices - weaving frame, hand looms with vertical, horizontal and circular arrangement of warp. The design of looms was determined by the type of material to be processed, the weave of the fabric, climatic conditions and the way of life of the people. Nowadays a modern weaving machine is high-speed, computerised, with excellent ergonomics, with a wide assortment of fabrics produced with quick change of assortment, with production of high quality fabrics. Almost all peoples of the world have myths and legends connected with fabric manufacturing, which were reflected in the literature and art of the time. An example of this is a verse treatment of the ancient Iranian story "Shahnameh" by Ferdowsi - for the dressing of silk, furs, cloth, from cocoons, skins and light linen, he taught to spin threads and stood behind the machine, to weave cunningly into the base of the duck. As we can see the art of decorating fabrics appeared in ancient times. Man has always strived to make his clothes dressy and comfortable, ie strived to create different patterns on the fabric. Pattern on the fabric can be obtained by weaving by interweaving warp and weft threads and in the process of finishing the finished fabric, embroidery or printing. Pattern formation by weaving is accompanied by artistic and technological process, the combination of which allows to achieve a variety of light and texture effects in the pattern, through different reflection of light from different parts of the fabric. The peculiarity and expressiveness of the pattern is ensured by interweaving different fabric structures and weaving them on a remise or jacquard loom.

CHAPTER 1

1. COMPLEX WEAVES

Main, derived and combined weaves are considered simple, as they are constructed with one main thread system and one weft system. Complex weaves are those in which several main thread systems and several weft systems are involved. Each of the thread systems is placed one above the other, forming layers of fabric. Therefore, the front and back sides of the fabric are independent weaves, which are called double-faced or two-faced weaves. Double-faced weaves are those that have the same front and back sides of the fabric. Double-faced weaves are those that have differences in the front and back sides of the fabric.

Advantages of double-faced and double-sided weaves:

- increase in fabric thickness and weight, without changing the linear density of warp and weft yarns;
- obtaining the same or different types of weaves on both sides of the fabric;
- formation on both sides of the fabric of the same or different yarns in colour, quality, type of yarn.

Filling and working out complex weaves is accompanied by some difficulties:

1.The participation of several warp and weft systems in the formation of a fabric requires several navaids and multi-colour (multi weft) looms.

2.In some cases, special weaving machines are required for specific fabrics such as weft woollen, warp woollen, buttonhole, piquet, openwork etc.

H.Arrangement of warp and weft yarns in the fabric in layers on top of each other, not next to each other.

Fabrics produced by complex weaves are intended for domestic and technical purposes. The basis for the construction of complex weaves are main, derived and combined weaves.

Construct complex weaves using the following principles:

1 The weft and main weaves of single-ply fabrics serve as a base;

2 .The long outer decks of the top layer should be placed closer together to cover the inner short overlaps of the lower fabric layer;

3 .Short internal slabs should be placed as far as possible in the middle of long decks;

4 The diagonals on the outside of the top layer and on the outside of the bottom layer point in opposite directions (based on twill weaves);

5 .When constructing double-sided weaves, the ratio between yarn systems can be 1:1, 1:2,2:1;

6 .When constructing double-faced weaves, the ratio between yarn systems can be 1:1;

7 .The weaves of the individual layers may be the same or different and coordinated in terms of the size of the rapport;

8 .The ratio of warp to weft density in plies shall be 1:1, 1:2, 2:1;

9 . The threads of the upper layer are labelled with Arabic numerals and the threads of the lower layer with Roman numerals.

Complex weaves are subdivided into the following types: -Half-ply double-sided and double-faced weaves;

- two-ply weaves: sack or hollow; double or multiple width fabrics; with ply movement weave; two-ply fabrics with different methods of ply dressing;
- of the weave of the piqué fabric;
- multilayer weaves;
- tufted fabrics, both duck-wool and basic-wool;
- ligature (openwork) fabrics;
- quilting (terry) fabrics.

CHAPTER 2

2. ONE-AND-A-HALF LAYER WEAVES

One and a half ply weaves are weaves that use two warp systems (with additional warp) and one weft system or two weft systems (with additional weft) and one warp system. In the first case, the warp systems are placed on top of each other and the weft threads are interwoven with the warp threads to join them together. In this case, the weft is subjected to high stresses and has a high workmanship. In the second case, the weft yarn systems are placed one above the other, and the warp yarns are interwoven with the weft yarns to make their connection. In this case, the warp is subjected to high stresses and has high workmanship.

In one-and-a-half-ply weaves, fabric thickness and weight can be increased without yarn thickening, and the same (double-faced) weave or different (double-sided) weaves can be obtained on the front and back sides.

These weaves are subdivided according to the method of construction:

1 .Double-faced and double-sided weaves with an additional warp. Formed from two warp systems tied with a common weft.

Weaves that have a long main overlap (basic twills, satin) are used for construction;

2 .Double-faced and double-sided weaves with an additional weft. It is formed from two weft systems tied together with a common backing. Weaves with long weft overlaps (weft twill, satins) are used for construction.

Methodology of construction of one and a half layer weaves

3 are determined by the weave of the outer side of the top layer and the outer side of the bottom layer and the ratio between the yarn systems.

4 .The base rapport of a one and a half ply weave with extra warp is equal to the product of the base rapport of the basic weave by the sum of the ratio of the main yarns, and with extra weft is equal to the base rapport of the basic weave.

5 The weft length of a weft pattern of a one-and-a-half-ply weave with additional warp is equal to the weft length of the basic weave, and with additional weft is equal to the product of the weft length of the basic weave by the sum of the ratio of weft yarns.

6 .The location of the inner short overlap is determined from the tissue section.

7 .Make up the weave of the inner side of the bottom layer, taking into account the position of the inner short overlap.

8 .Make up a complete filling pattern for a one and a half layer fabric.

Construction of a one and a half layer double-faced fabric with an additional backing

The weave on the outside of the top layer is basic twill 3/1 (Fig. 1a). The weave on the outside of the bottom layer is basic twill 3/1 (Fig. 1b). The ratio between the warp systems is 1:1. The warp and weft ratio of the basic weave for the top layer $Ro_B = R_{(yB)}) = 4$ (Fig.1a), and for the bottom layer $R_{0H} = R(yH)) = 4$ (Fig.1b). According to point 9 of the principle of construction of complex weaves, the warp yarns of the upper layer are labelled with Arabic numerals and the warp yarns of the lower layer with Roman numerals, as the weft yarns remain labelled with Arabic numerals. The weave of the inner side of the warp yarns of the bottom layer with the weft yarns is shown in (Fig. 1c). The warp pattern Ro = 4 (1+1) = 8, and the weft pattern $R_y = 4$.

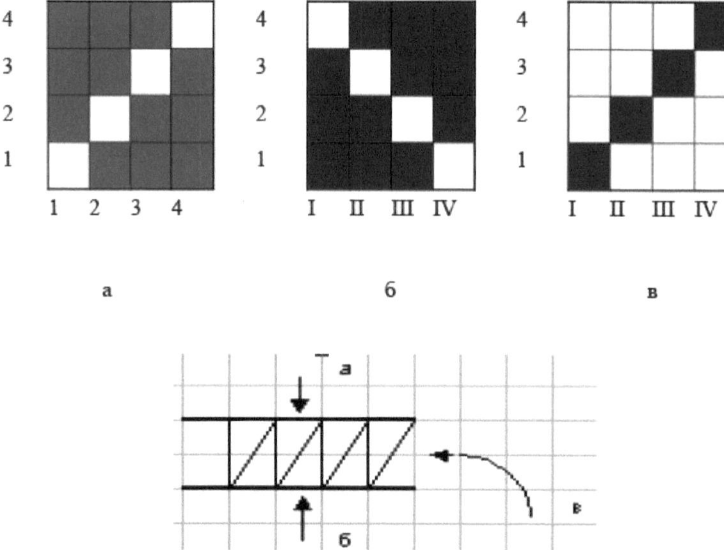

Fig. 1. Weave of the upper and lower layers of fabric: a - weave on the outside of the upper layer of fabric; b - weave on the outside of the lower layer of fabric; c - weave on the inside of the fabric of warp threads of the lower layer with weft threads.

Let us show the diagram of the longitudinal section of the fabric on the first warp (Fig. 2a) of the upper warp (1) and determine the location of the short inner overlap. In this case (according to point 3 of the principle of construction of complex weaves) it is reasonable to locate the inner short overlap on the third weft thread, i.e. the first warp thread of the bottom layer will be the third thread of the bottom warp of the inner side weave (Fig.1c and 2d). By successively cutting the fabric for the second and subsequent warp yarns we determine the location of the short inner overlap for the other warp yarns of the bottom layer of the fabric (Fig.1c and 26, c, d, D.).

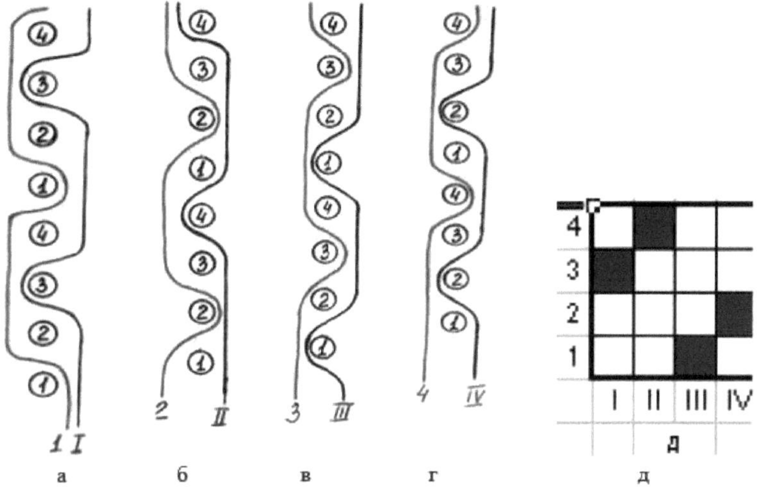

Fig. 2. Schematic of the longitudinal section of the fabric (a, b, c, d) and the weave of short internal overlaps in the fabric (e) of a one and a half layer double-faced fabric with additional backing

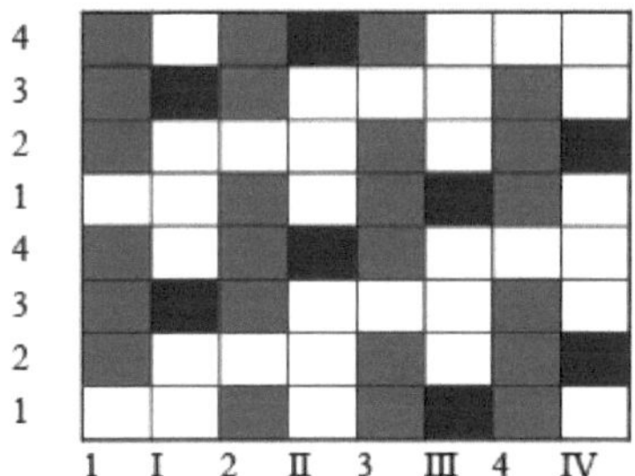

Figure 3. Filling pattern of one and a half layer double-faced fabric with additional basis

Draw up the filling pattern of a one and a half layer double-faced fabric with an additional warp (Fig. H). To do this, we transfer the weave pattern of the upper warp layer (Fig.1a) and then the weave of the short inner overlaps of the lower warp layer (Fig.1c and 2d). The remise can be either row or vaulted, preferably vaulted, as the first vault is used for the less stressed warps of the bottom layer. The number of remizas in the dressing is equal to the sum of rapports on the base of the base weave of the lower and upper layer $K = R_{OB} + R_{OH}$

The reed tooth is threaded with a number of threads equal to or multiple of the sum of the ratios of the warp systems.

Construction of a one and a half layer double-sided fabric with an extra backing.

The weave on the outside of the upper layer of the fabric is corduroy 2/2 (Fig.4a). The weave on the outside of the bottom layer is four-thread satin (Fig. 4b). The ratio between warp systems is 1:1.

Ply by ply, complex weaves $R_{O_B} = R_{OH} = R_{(yB)} = R_{yH} = 4$

Half-layer fabric base pattern $R_O = 4(1 + 1) = 8$

$R_y = 4$

We take a longitudinal section of the fabric on the first warp yarn of the top layer (Fig. 4g) and determine the location of the short inner overlap. In this case, the short inner overlap is located on the first weft yarn. Fig.4c shows the weave of the inner side of the warp yarns of the bottom layer with the weft yarns. Then we transfer Figs. 4a and 4c to the filling pattern 4d of a one and a half ply double-sided fabric with an additional warp.

Filling and production parameters are similar to the previous weave built on the basis of twill 3/1.

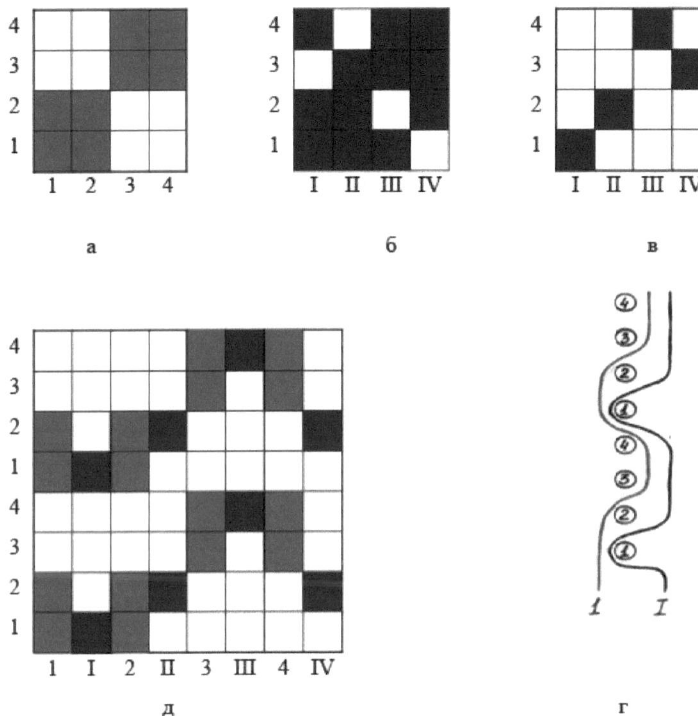

Fig. 4.Weave One and a half ply double-sided fabric with
a - weave on the outer side of the upper fabric layer; b - weave on the outer side of the lower fabric layer; c - weave on the inner side of the fabric of warp yarns of the lower layer with weft yarns; d - longitudinal cut of the fabric along the first warp yarn of the upper layer; e - filling pattern of one and a half layer double-sided fabric with additional backing.

Construction of a one and a half layer double-faced fabric with an additional weft

weft.

Weave on the outside of the top and bottom layers is 1/3 weft twill. The ratio between weft yarn systems is 1:1. The ratio of the basic weave on the warp and weft of the upper and lower layers is $R_{OB} = R_{OH} = R_{(yB)} = Ry_H = 4$. The upper weft yarns are labelled with Arabic numerals and the lower weft yarns with Roman numerals (Fig. 5a, 56). The weave of the inner side of the weft yarns of the lower layer with the warp yarns is shown in Fig. 5c.

a б в

Fig. 5. Weave of the upper and lower layers of a one and a half layer double-faced fabric with additional weft: a - weave on the outer side of the upper fabric layer; b - weave on the outer side of the lower fabric layer; c - weave on the inner side of the fabric of the weft threads of the lower layer with warp threads.

Ramp of one and a half layer double-faced fabric with additional weft on the weft $R_y = 4\ (1 + 1) = 8$

Ramp of one and a half ply double-faced fabric with additional fabric weft on the base $R_o = 4$.

Here is a diagram of a cross section of a one and a half layer double-faced fabric with additional weft on the first weft thread (Fig. ba) and determine the location of the short inner overlap of the bottom layer.

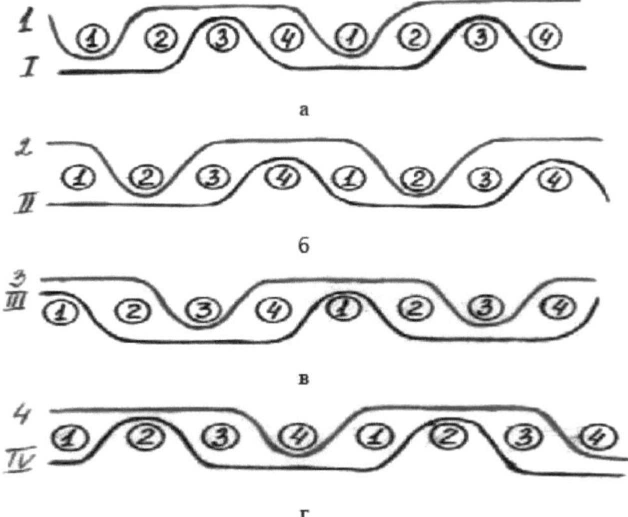

Fig. 6. Schematic diagram of a cross section of a one and a half layer double-faced fabric with additional weft: a - along the first weft thread; b - along the second weft thread; c - along the third weft thread; d - along the fourth weft thread.

In this case, it is reasonable to place the inner short overlaps on the third main yarn. Therefore, the first weft thread of the bottom layer will be the third weft thread of the inner weave (Fig. 5c), and the second weft thread will be the fourth, the third - the first, the fourth - the second. The number of wefts is equal to the base rapport k R =

Construction of a one and a half layer double-sided fabric with additional weft

weft.

The weave of the outside of the top layer is satin 5/2 and the bottom layer is twill 1/4. The ratio of the weft yarn of the upper layer to the lower layer is 2:1. Rapport of the basic weave on the warp and weft of the upper and lower layers of the fabric (Fig.7a and 76) ROB= ROH= RyB= Ry_H= 5. k = RO.

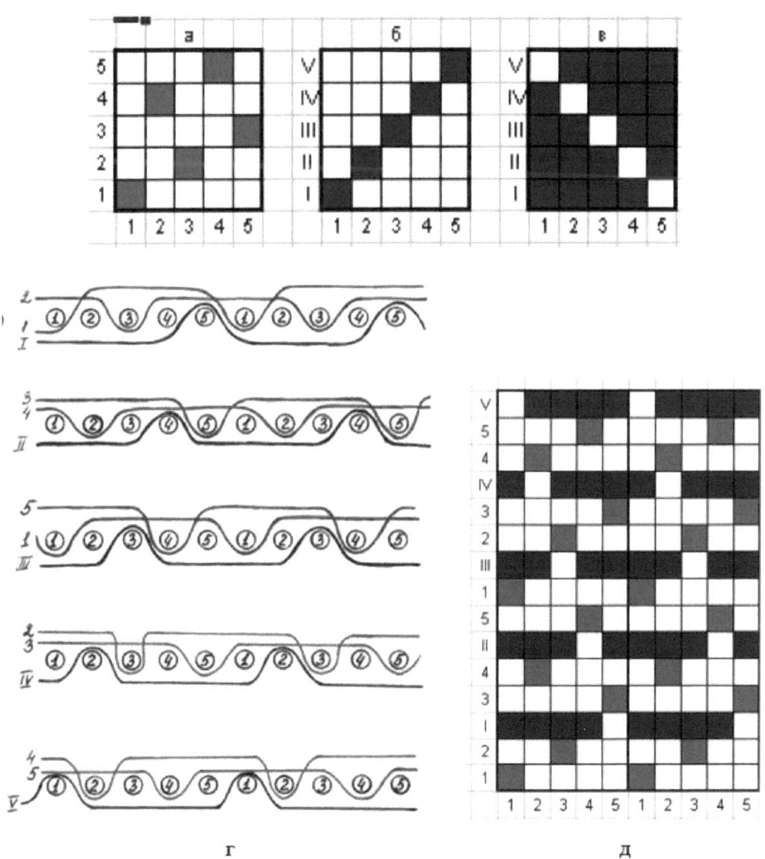

г д

Fig.7.Weave of one and a half layer double-sided fabric with additional weft: a - weave on the outside of the upper layer of fabric; b - weave on the outside of the lower layer of fabric; c - weave on the inside of the fabric of weft threads of the lower layer with warp threads; d - cross section of fabric; e - filling pattern of one and a half layer double-sided fabric with additional weft.

$R_y = 5\ (2 + 1) = 15$. The pattern of one and a half layer double-sided fabric on the base Ro = 5. Fig. 7c shows the weave of the inner side of the weft yarns of the bottom layer with the warp yarns, and Fig. 7d cross section of the fabric and in Fig. 7d the tucking pattern of a one and a half ply double sided fabric. The purling in the remise is rowwise, the number of remise is equal to the warp rapport.

The samples of one and a half ply fabrics are analysed in the same way as one and a half ply fabrics, taking into account their structure. When determining the front and back sides, it is taken into account that the front side is produced from higher quality (expensive) raw materials and the direction of diagonals from left to right upwards, as well as has a pronounced weave.

The warp and weft directions are determined by the following characteristics:

1 .When there is an edge, the direction of the base and edge are the same;

2 .The linear density of the main yarns is less than the linear density of the weft yarns;

3 .The main threads have a lot of twist;

4.According to the specific features of the fabrics (the beard's tooth, twins, undercuts, etc.).

The warp and weft yarn processing is determined by the difference between the lengths of the straightened yarns and the length of the fabric, both for the upper and lower layers of the fabric. When determining the weave of the outer side of the upper layer fabric with additional weft, the lower weft yarns are removed from the sample, and for fabrics with additional warp, the lower warp yarns are removed. To do this, form a fringe at the bottom and carefully remove the corresponding threads from the left side of the sample by the tips. The weave pattern is then transferred to the canvass paper.

When determining the weave of the outer side of a bottom layer fabric with additional weft, the upper weft yarns are removed from the sample, and for fabrics with additional warp, the upper warp yarns are removed. Then, the weave pattern is drawn on canvass paper. The ratio of the yarn systems in the fabric layers is then determined by counting.

The weave and yarn ratio in the layers determine the weave pattern of a one and a half ply fabric in warp and weft.

When determining the weave of the inner side of the lower fabric layer, two weft yarns (fabrics with extra weft) or two main yarns (fabrics with extra warp) are brought out on the fringe and their interposition is revealed . After depicting the weave pattern of a one-and-a-half layer fabric, the number of fringes in the dressing and the type of the main threads in the reed and fringes are determined.

Conclusion

One and a half ply fabrics are produced on weaving machines equipped with dobby shedding machines, multi-colour (multi needle) machines and double warp threading (for warp yarns with different processing). The threading of double-faced and double-sided fabrics with additional warp causes an increase in the number of heddles and the consolidated selvedge makes the selvedging process more complicated. The warp threads lying one above the other are picked in one reed tooth. Double-faced and double-sided fabrics with additional weft do not require an increase in the number of reams, but the weaving machine productivity in metres is considerably reduced A= n - 60/Ru -10, m/hour, where: n is the machine main shaft speed, min^{-1}, Ru is the weft density of the fabric, yarns/dm.

The selvedges of one and a half ply fabrics have a rep weave and use yarns with high elastic properties (kapron, etc.).

CHAPTER 3

3. DOUBLE-LAYER WEAVES

The peculiarity of double-layer fabrics is that two independent warp and two independent weft systems are involved in their construction. At the same time they form two independent layers lying on top of each other, free or connected to each other during the weaving process.

Two-ply weaves are subdivided according to the way the individual layers are tied:

- baggy or hollow weaves;
- weaves to form a fabric of multiple widths;
- weaves to form a fabric with moving layers;
- weaves of two-layer fabrics with different methods of ply tying.

Sack or hollow weaves

They are used for the production of fire or field hoses, technical cloth, bags. The basic weaves are plain weave, corduroy 2/2, weft 2/2, four-strand sateen. The following principles are used in the formation of sack weaves:

I.Hπτπ warp and weft threads of the upper fabric layer are numbered with Arabic numerals, and warp and weft threads of the lower fabric layer with Roman numerals.

2 .When depicting a weave on paper, the warp and weft yarns are conventionally shifted to the same plane.

3 .When weft is added to the upper layer, all warp threads of the lower fabric layer are dropped (pnc.la).

4 .When weft is placed in the lower fabric layer, all warp threads of the upper fabric layer are lifted (Fig. 1b).

5 The lifting of the warp yarns of the upper fabric layer when the weft is inserted into the lower fabric layer is shown in the drawing as a circle.

6 The fabric layers are joined at the edges of the dressing by alternating the wefts.

7 The warp and weft rapport (R) of a fabric is equal to the smallest multiple of the number of yarns in the base weave rapport (Re) multiplied by the number of fabric layers (K).

$$R= Re -K$$

8 .Bass cords are lowered when forming the upper layer of the hollow fabric, and raised when forming the lower layer of the fabric (Fig.2)

Fig. 1 Scheme of formation of two-layer hollow tissues

To maintain the established density of the hollow fabric, bass cords are used in the places of transition of the weft from one layer to another, i.e. in the places of folds of the hollow fabric. The cords are threaded into individual wefts and reed tines.

The bass cords are lowered during the formation of the upper fabric layer and raised during the formation of the lower fabric layer. This means that they are not worked by the weft (Fig. 2) and can be freely removed after removing the fabric from the loom.

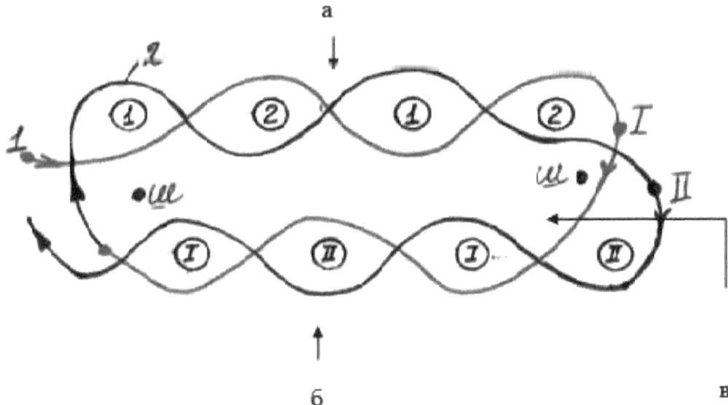

Fig. 2. View of a two-layer hollow tissue: a - from above; b - from below; c - inside.

Let's construct the filling pattern of hollow fabric on the basis of plain weave, the rapport of basic weave Re = 2, and the ratio of fabric layers let's assume 1 : 1 (Fig.Za and 36).

Hollow fabric warp and weft ratio

$$Ro= {}_{R(y)}= 2Re = 4$$

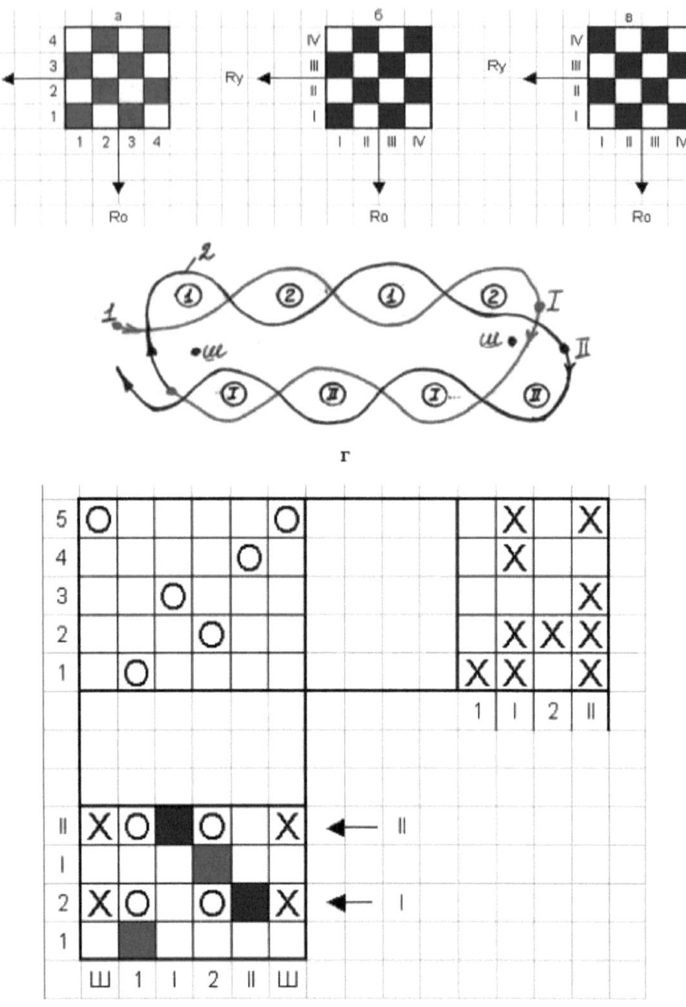

Fig.Z. Weave of two-layer hollow fabric: a - weave on the outside of the upper layer of fabric; b - weave on the outside of the lower layer of fabric; c - internal weave of the lower layer of hollow fabric; d - cross section of hollow fabric; e - filling pattern of two-layer hollow fabric.

We draw the cross section of the fabric (Fig. Zg) and define the internal weave of the bottom layer of the fabric (Fig. 2 and Zc). Then transfer the weave Za and Zv to the filling pattern Zd.

Using the principles of points 3, 4, 5 and 8 of hollow fabric construction, draw up a complete filling pattern for a hollow fabric. The alternation of weft skeins takes place in the following sequence:

- first weft throw for the top layer of fabric;
- A second skip for the bottom layer of fabric;
- A third skip for the top layer of fabric;
- The fourth skip is for the bottom layer of fabric.

The bass cords (B) are lifted from the individual loom when the bottom layer of the fabric is formed. The process of removing bass cords after removing the hollow fabric from the loom is very labour-intensive. Therefore, it is expedient to develop means of maintaining a given density of the fabric in the places of folding (transition of wefts from one layer to another layer), which would replace bass cords on the machine. One of the variants of arrangement between the layers in the fabric at the fabric edge is a strip (P) - rectangular or trapezoidal, oval and other shapes (Fig.1).

Double width fabric weaves

They are developed when fabrics of greater width than the loom width allows are required. These weaves are limited in their application for the following reasons: special wide looms are available; machine threading is very difficult; machine output in metres is reduced.

When producing double width fabrics, the weft yarn laying sequence can be as follows:

1 .First weft laying for the top fabric layer, second and third for the bottom fabric layer and fourth for the top fabric layer;

2 .First and second weft laying for the upper fabric layer, third and fourth weft laying for the lower fabric layer.

Double width fabrics are constructed in the same way as hollow fabrics.

Let's build the filling pattern of double width fabric on the basis of plain weave, the basic weave pattern Re = 2, and the ratio of the fabric layers is assumed to be 1:1. Hollow fabric pattern on warp and weft

$$Ro= {}_{R(y)} =2Re = 4.$$

We will adopt the second method of weft insertion, i.e. two insertions of weft threads in the upper fabric layer and two insertions of weft threads in the lower fabric layer.

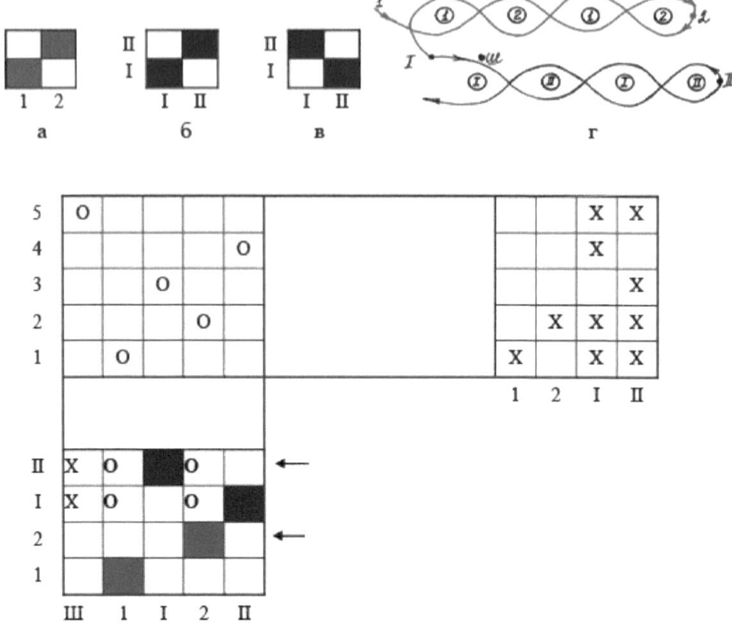

Fig.4.Weave of double-layer double-width fabric: a - weave on the outside of the upper fabric layer; b - weave on the outside of the lower fabric layer; c - inner weave of the lower fabric layer; d - cross section of the fabric; e - complete filling pattern of double-layer double-width fabric.

Draw the cross section of the fabric (Fig. 4d) and make the weave of the inner side of the fabric (Fig. 4c). Then, we transfer the weave of the upper layer of fabric (Fig. 4a) and the weave of the inner side of the fabric (Fig. 4c) to the filling pattern of the fabric (Fig. 4d). Taking into account the rules of construction of double width fabrics, we make a complete filling pattern (Fig. 4d).

Weaves of fabric of multiple widths.

Peculiarities of construction of a fabric of multiple widths:

1 The warp and weft yarns of the layers are placed side by side in one plane in the weave pattern.

2 The warp and weft threads of the first and second layers are designated respectively by Arabic and Roman numerals, and the subsequent layers (third, fourth, etc.) by the letters of the alphabet a, b, v, etc.

3 .The warp yarns of odd-numbered layers show the outside of the base weave, while even-numbered layers show the inside of the base weave .

4 .When weft is laid in the top layer, some of the warp yarns of the top layer (according to the weave pattern) are raised and the rest of the warp yarns of the underlying layers are lowered.
5 .When weft is placed in the bottom layer, all warp and bass cords of the layers above are lifted, as well as the warp cords of this layer according to the weave pattern.
6 .The weave pattern of multiple width fabrics is equal to the product of the basic pattern (Kb) by the number of layers (K), R = Kb - K.

Conclusion

The production of hollow, double and multiple width fabrics is carried out on special weaving machines equipped with shedding dobby or eccentric mechanisms. The number of reams is determined as the sum of rapports on the basis of the base weaves of the layers and the number of reams is equal to the number of fabric layers. The number of warp threads is equal or multiple to the number of layers in the fabric. These fabrics are used for fire hoses, transport tapes, bags without seam, and other woven products.

CHAPTER 4

4.TWO-LAYER WEAVES WITH DISPLACEMENT OF LAYERS IN FABRIC

The characteristic feature of a weave with layer movement is that two layers of fabric are joined together along the contour of a pattern by moving the layers. The pattern can be longitudinal and transverse strips, squares, checkers, etc. In this case, closed-hollow (capillary-shaped) relief bilateral patterns (fabric sections) are formed, differing from each other by the colour of warp and weft threads or their type (thickness, quality, etc.). To produce fabrics of such weaves, two warp and two weft systems are required and the ratio between the yarn systems is 1:1, and if yarns of different linear densities are used, 2:1 or 1:2 is possible. The basic weaves used are plain weave, twill, cambric 2/2 and others. The weave ratio depends on the size of the pattern motif (width of the plot), the ratio between the thread systems in the fabric layers, the density of the fabric in the warp and weft, and the basic weave ratio. The construction of the weave of these fabrics is based on the construction of hollow fabrics. However, at the boundary of the fabric pattern, the yarns in the fabric layers change position, i.e., the warp and weft yarns of the upper layer move to the lower fabric layer, and the warp and weft yarns of the lower layer move to the upper fabric layer. Due to this movement of warp and weft yarns in the plies, the plies of the fabric are joined together. Fig. 1 shows a weave with displacement of layers, where it can be seen in the places of transition of threads from layer to layer between warp threads 8-9 and 16-1 and between weft threads V111-1X and XV1-1 a depression is formed, as a result of which the squares are convex. To build a weave with moving layers it is necessary to have the pattern motif and its dimensions, the basic weave, the density of the fabric on the warp and weft. Let's build a filling pattern of the fabric with moving layers along the contour of the pattern. The pattern is represented as black and white rectangles with six warp and three weft threads in each layer (Fig.2). The ratio of layers is 1:1, the base weave is twill 1/2.

Figure 1. View and section of a two-layer tissue with displacement of layers.

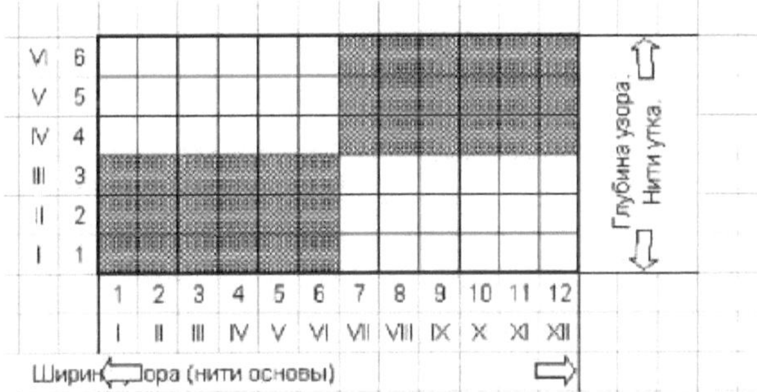

Figure 2. Pattern motif of a two-layer fabric with moving layers.

Top layer base rapport $Ro_B = 6 + 6 = 12$

Upper weft pattern $R_{yB} = 3 + 3 = 6$

Fabric spread on the base $Ro = 2\text{-} Ro_B = 2\text{-} 12 = 24$

$R_y = 2\text{-} R_{yB} = 2\text{-} 6 = 12$

Let us represent the weave of the outer side of the top layer (Fig. Za) and the inner side of the bottom layer (Fig. Zb). Transfer the weave (Fig. Z and Fig. Zb) to the filling pattern (Fig. Zc). Draw the upper layer warp lifts (O) for the interlacing of the warp and weft yarns of the lower layer.

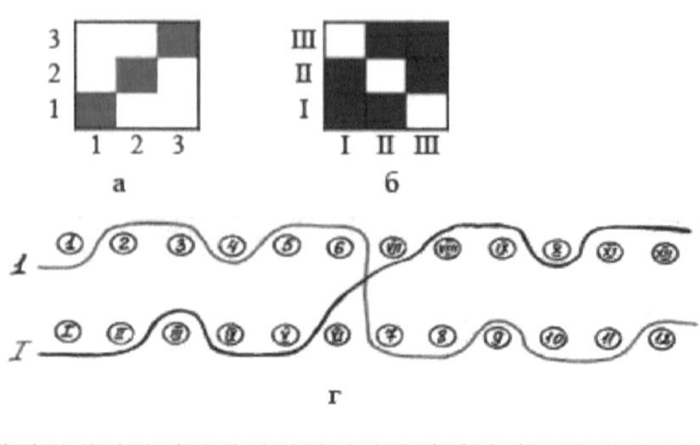

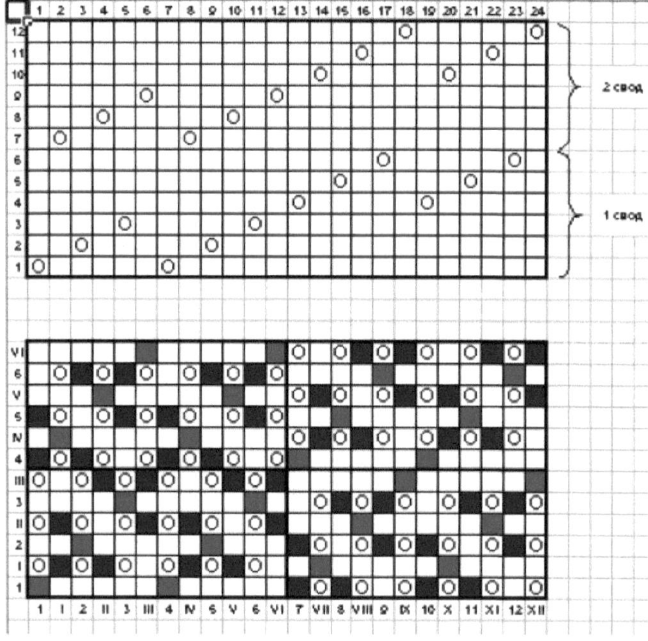

в

Fig.Z. Weave of two-layer fabric with moving layers along the pattern contour: a - weave on the outside of the upper fabric layer; b - weave on the outside of the lower fabric layer; c - inner weave of the lower fabric layer; d - cross section of the fabric; e - tucking pattern of two-layer fabric with moving layers along the pattern contour.

It should be noted that the warp and weft of the same colour are either upper or lower along the length of the rapport. If in the first pattern element they were upper, in the second pattern element they will be lower, which is clearly illustrated by the cross section of the fabric, and the joining of layers takes place after six warp threads of the upper and lower layers and after three weft threads of the upper and lower layers. The warp threads are interrupted along the vaults.

Conclusion

Two-ply fabrics with ply movement are produced on weaving machines equipped with dobbies or Jacquard machines, multicolour devices or multifilament weaving machines.

The warp threads are usually picked into the heddles, with warp threads of the same colour or type of raw material being picked into each heddle. The number of heddles corresponds to the type of basic weave, pattern motif and warp density of the fabric. The number of warp threads in the reed tooth is even (upper layer thread and lower layer thread). Base weaves can be plain weave, twill weave, etc. Two-ply fabrics with displaced layers are used as decorative, dress, coat, tablecloths and napkins.

5. TWO-PLY WEAVES WITH DIFFERENT PLY BINDING METHODS

Fabrics formed with this type of weave consist of two layers closely interlocked to form a single, double-thickness fabric. This type of weave is widely used in the production of winter drapes and weighted fabrics. The warp threads may be the same as the weft threads (thickness, type of fibre, etc.), but may be different depending on the nature of the fabric. The methods of tying (binding) the layers in two-ply fabrics are as follows:

1.Bottom-to-top dressing in double-layer fabrics, i.e. the bottom warp is tied to the top weft (Fig. 1).

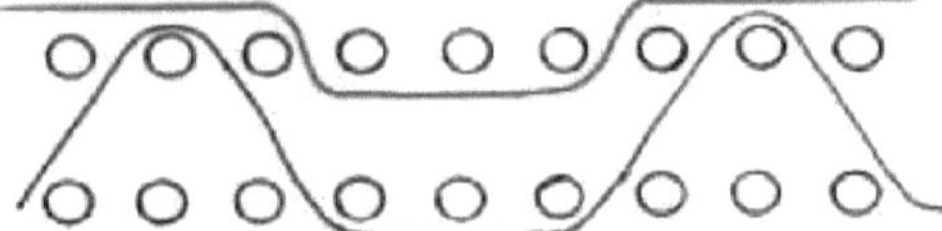

Figure 1. Layer dressing from bottom to top in bilayer tissues.

2.Top-to-bottom binding in two-ply fabrics, i.e. the upper warp yarns are tied to the lower weft yarns (Fig. 2).

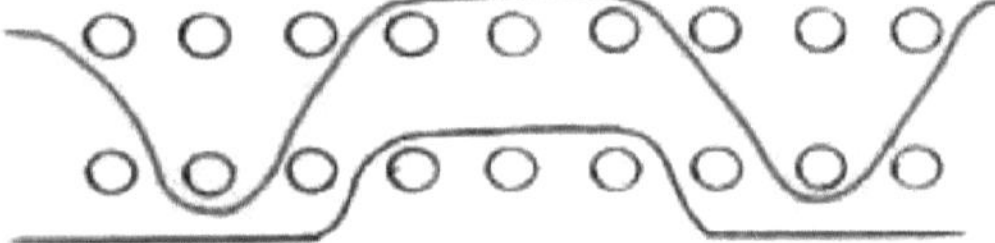

Figure 2. Layer dressing from top to bottom in bilayer fabrics.

3.Combined dressing in double-layer fabrics, i.e. the lower warp is tied to the upper weft and the upper warp to the lower weft (Fig.H).

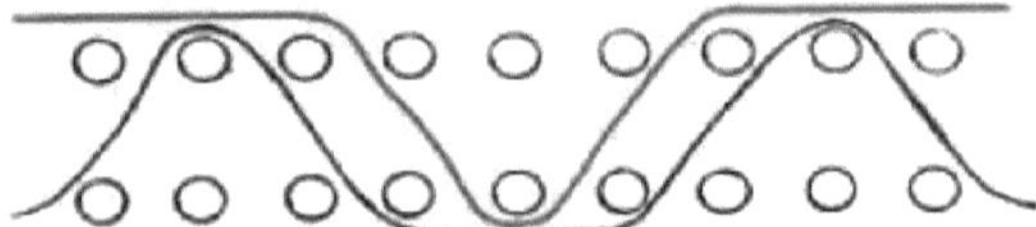

Figure 3. Combined layer dressing in bilayer tissues.

4.Binding with additional presser yarns, i.e. joining the layers with presser yarns of the weft (Fig.4) or presser yarns of the warp (Fig.5).

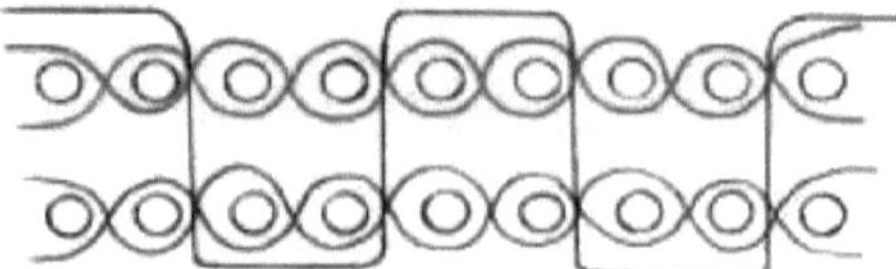

Figure *4*. Layer dressing with additional pressed weft yarns in two-ply fabrics.

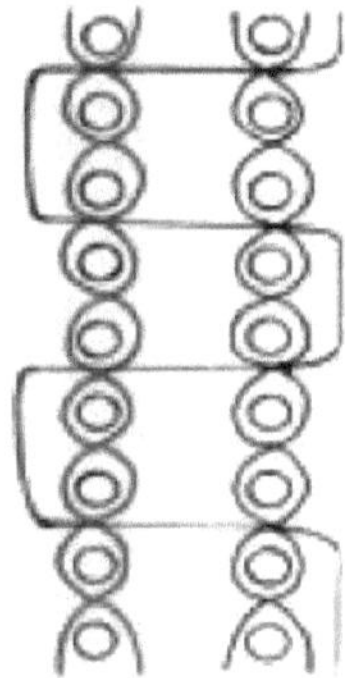

Figure 5. Layer dressing with additional pressing yarns in two-ply fabrics.

The method of dressing is established based on the following considerations:

- from the ratio of yarn densities between the layers;
- from the ratio of thread thickness, quality and colour between layers;
- from the nature of the purpose and weave of the front and back of the fabric.

When selecting the binding method, it is essential that the layers are firmly bound and single bindings must be hidden and distributed evenly within the fabric rapport. To construct these weaves, the base weaves are the fundamental and derived weaves, and both sides may be equally or unequally interlaced. The warp (Ro) and weft (R_y) ratio of a two-ply weave is equal to twice the smallest multiple of the number of threads of the basic weave (Re)

$$Ro= Ry= 2R$$

Two-ply weaves with bottom-to-top interlacing

The basic weave of the layers is twill 3/3, the ratio between the number of threads in the layers is 1:1. The weave is shown on the outside of the upper fabric layer in Fig. ba and on the inside of the lower fabric layer in Fig. 66. The warp and weft ratio of the basic weave is Re = 6. The warp and weft ratio of the two-layer fabrics

$$Ro= Ry 2\text{-}Re= = 2\text{-}6 = 12.$$

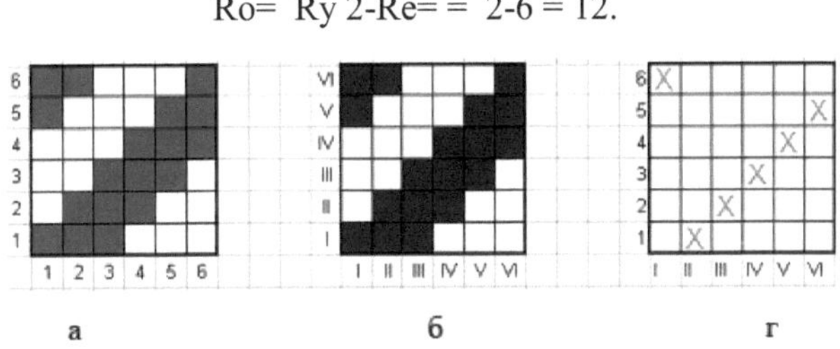

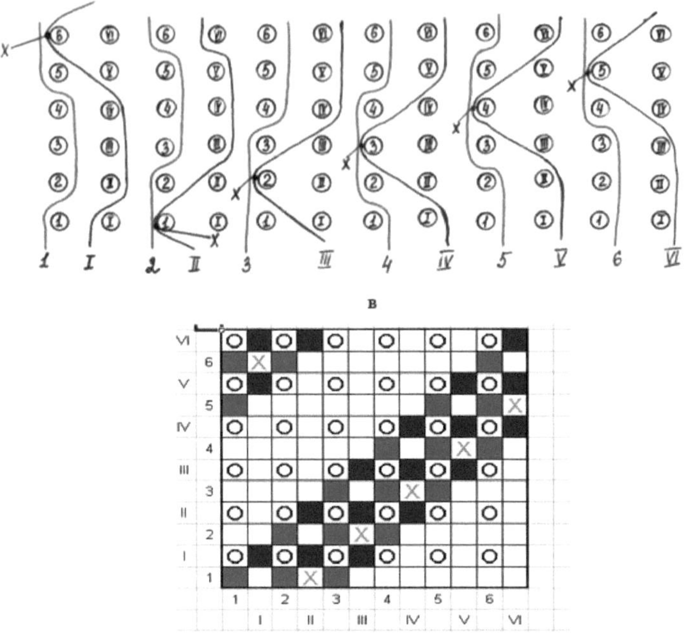

д

Fig. 6. Weave of a two-layer fabric with the connection of layers from bottom to top: a - weave from the outside of the upper layer of fabric; b - weave from the outside of the lower layer of fabric; c - longitudinal section of fabric and places of tying of layers by warp threads of the lower layer with weft threads of the upper layer; d - programme of tying of layers in fabric by warp threads of the lower layer with weft threads of the upper layer; e - filling pattern of a two-layer fabric with the connection of layers from bottom to top.

Fig. bc shows a longitudinal section of the fabric to determine the location of the short inner overlap of the bottom warp and top weft, using the principles of construction of a two-layer fabric. For the first warp thread of the lower fabric layer it will be the sixth weft of the upper fabric layer, the second warp thread - the first weft, and the third warp thread - the second weft, etc. The places where the bottom warp is tied to the upper weft are labelled (X), which means that the bottom warp is lifted to tie to the upper weft.

According to the longitudinal section of the fabric (Fig. bb), having determined the places of tying, we can make a programme of tying the layers (Fig. bg). Next, let's make a dressing pattern (Fig. bd) of a two-layer fabric with tying of layers from bottom to top. For this purpose we transfer the weave of the upper layer (Fig.ba) and the weave of the lower layer (Fig.bb) and the programme of layer dressing (Fig.bg). In Fig. bd and mark the lifts (O) of the upper warp layer while laying the lower weft. The purlins of the warp yarns in the remise are summarised and the number of remises in the filling (K) is determined by

$$K = R_{OB} + R_{OH} = 6 + 6 = 12$$

Two-ply weaves with top-to-bottom interlacing

The basic weave of twill layers is 3/3, the ratio between the number of threads in the layers is 1:1. Weave on the outside of the fabric layer Fig. 7a and on the inside of the bottom fabric layer Fig. 7b.

The warp and weft ratio of the basic weave is Re = 6.

Double-layer warp and weft lengths

$$R_O = Ry\ 2\text{-}Rö = = 2\text{-}6 = 12$$

Fig. 7c. shows a longitudinal section of the fabric, in which we use the principle of two-layer fabric construction to determine the short inner overlap of the upper warp and lower weft. It is reasonable to tie the first warp thread of the upper layer to the third weft thread of the lower layer, the second warp thread of the upper layer to the fourth weft thread of the lower layer, and so on. (Fig. 7c). The places where the upper warp is tied to the lower weft are marked (□), which means that the upper warp is lowered to tie to the lower weft.

According to the longitudinal section of the fabric (Fig. 7c), having determined the places of layer dressing we can make a programme of layer dressing (Fig. 7d).

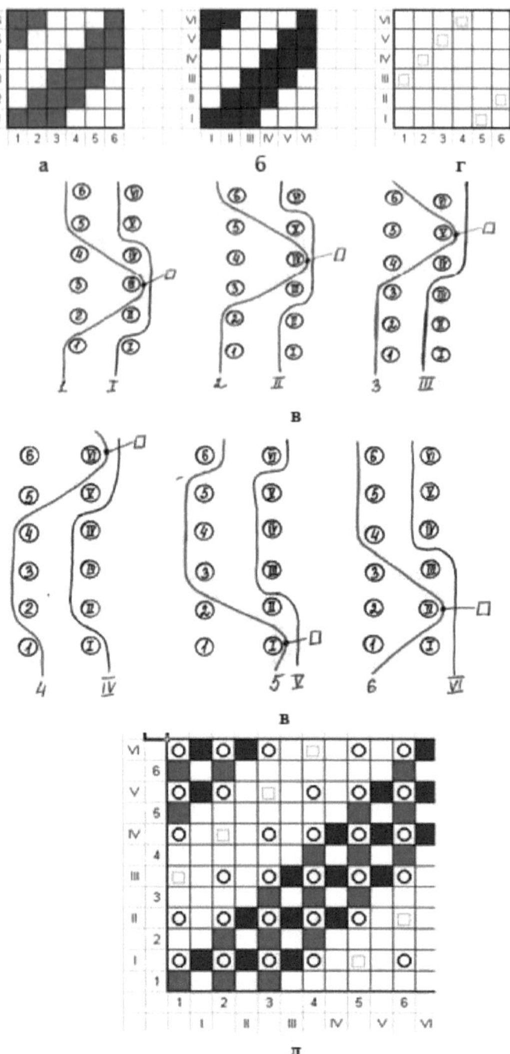

Figure 7. Weave of a two-layer fabric with layers joined from top to bottom: a - weave on the outside of the upper layer of fabric; b - weave on the outside of the lower layer of fabric; c - longitudinal section of fabric and places of tying of layers by warp yarns of the upper layer with weft yarns of the lower layer; d - programme of tying of fabric layers by warp yarns of the upper layer with weft yarns of the lower layer; e - tucking pattern of two-layer fabric with joining of layers from top to bottom.

On the filling pattern (Fig.7d) we transfer the weave of the upper layer (Fig.7a), the weave of the lower layer (Fig.7b), the programme of layer binding (Fig.7d) and mark the lifting of the warp yarn of the upper layer when forming the lower layer of the fabric. Selection of warp threads in the remise is summarised for 12 remises.

Two-ply weaves with combined ply binding

This weave is used to produce a strong, dense fabric. This is possible by tying the upper warp with the lower weft and the lower warp with the upper weft. In a combined weave, bottom-up and top-down weaves are used at the same time. Consequently, the rules of weave pattern construction of a two-layer fabric with combined binding are a combination of the rules of "bottom-up" and "top-down" binding stated above. Let's take the basic twill weave 3/3 ratio of the number of threads in the layers 1:1.

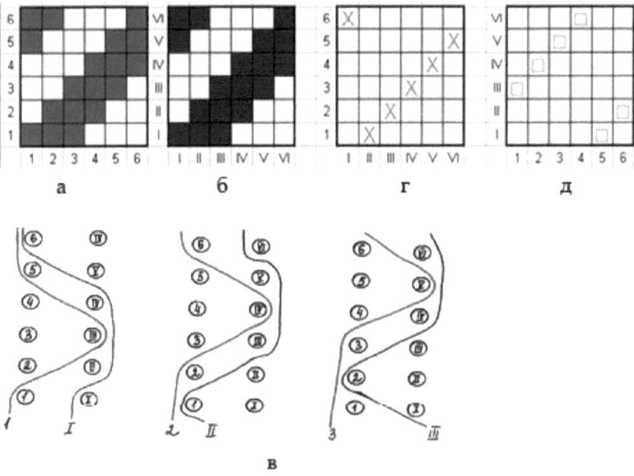

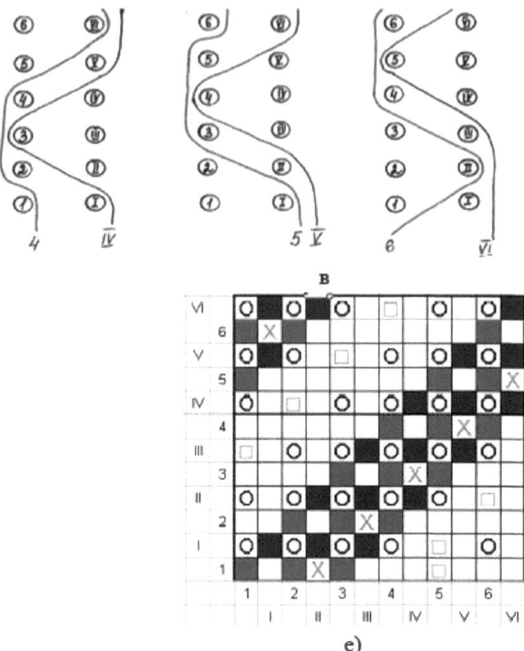

e)

Figure 8. Weave of a two-layer fabric with combined binding of layers: a - weave on the outer side of the upper fabric layer; b - weave on the outer side of the lower fabric layer; c - longitudinal section of the fabric and places of dressing of the upper warp layers with lower weft and lower warp layers with upper weft; d - programme of dressing of fabric layers by warp yarns of the lower layer with weft yarns of the upper layer; e - programme of dressing of fabric layers by warp yarns of the upper layer with weft yarns of the lower layer; f - filling pattern of two-layer fabric with combined dressing of layers.

The length of the two-layer fabric with combined warp and weft binding is 12 threads. The weave on the outer side of the upper fabric layer in Fig. 8a and on the inner side of the lower fabric layer in Fig. 8b is presented. According to the longitudinal section (Fig.8c) we will determine the places of binding of the upper warp with the lower weft and the lower warp with the upper weft without violating the principles of construction of complex weaves, and make a programme of binding of the lower warp with the upper weft (Fig.8d), as well as of the upper warp with the lower weft.

weft (Fig. 8d). Then we transfer the obtained weaves (Fig.8 - a, b, d, e) to the dressing pattern of the two-layer fabric with combined dressing (Fig.8e), taking into account the rules of construction of these fabrics. The purl in remiz is consolidated for 12 remiz.

Two-ply weaves with pressed backing dressings

The special feature is that each presser warp yarn is tied to one weft yarn of the upper and lower layers and three warp systems (upper, lower and presser) and two weft systems (upper and lower layers) are required. The ratio between the yarn systems can be varied. The weave ratio is determined by the product of the smallest multiple of the base ply weaves by the sum of the yarn system ratios. This takes into account the weave ratio of the pressed warp with the weft yarns of the upper and lower layers. The pressed warp has a high workmanship, so it is wound on a separate warp. A three-strand warp is taken in the remiz and the presser, upper and lower warp yarns are taken into the reed tooth. Let's build a filling pattern of a two-layer fabric with a pressed warp on the basis of the basic twill weave 2/2, the ratio of warp systems 1:1:1 and 1:1 for weft yarns. The weft fabric rapport: $R_y = (1+1)\ \text{-Kb}^{=}\ (1+1)\ '4 = 8$.

Rape of the fabric on the base: Ro = (1+1+1) -Kb = (1+1+1) -4 = 12.

The number of heddles in the dressing is 12, the threads in the heddle are three-stranded and there are three warp threads (upper, lower and presser) in the reed tooth. Figures 9a and 96 show the weaves of the outer and inner sides of the fabric. By depicting the longitudinal section of Fig.9c, we determine that several variations of the presser warp binding with the upper weft and the lower weft are possible. For the first longitudinal section, these can be the following tying of the presser backing: with the third lower and the first upper weft; with the second lower and the fourth upper weft; with the second lower and the first upper weft; with the third lower and the fourth upper weft.

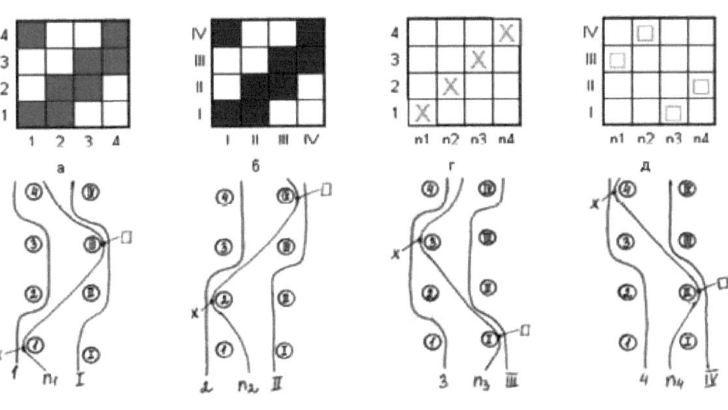

B

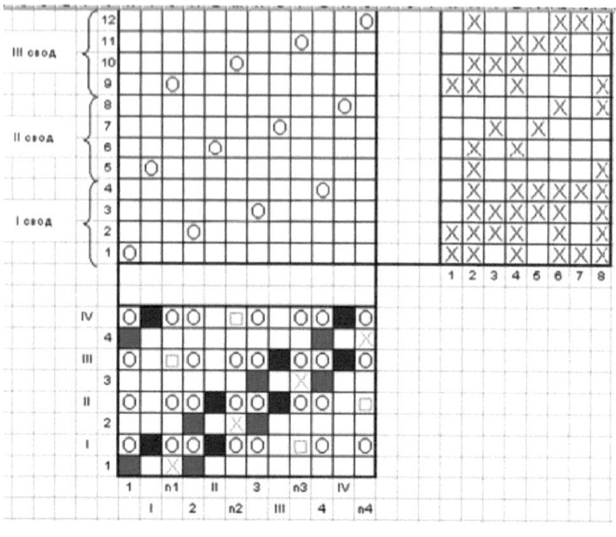

e

Figure 9. Weaving of a two-layer fabric by tying the layers with a pressing base: a - weave on the outside of the upper fabric layer; b - weave on the outside of the lower fabric layer; c - longitudinal section of the fabric and places of tying the layers by the pressure warp with upper weft and lower weft; d - programme for tying the layers in the fabric with the pressure warp yarns with upper weft and lower weft; e - programme for tying the fabric layers with the pressure warp yarns with upper weft and lower weft; f - filling pattern of the two-layer fabric by tying the layers with the pressure warp yarns.

The section shows a variant of tying the clamping base with the first upper and the third lower sinker. As can be seen from the section, when tied with the upper wefts, the presser backing has an upward rise, and when tied with the lower wefts, it has a downward fall. Let's denote the places of dressing by the lifting of the pressed warp above the upper weft (X), and the lowering of the pressed warp under the lower weft (□) Let's make a programme of dressing of the pressed warp **(i)** with the weft threads of the upper layer (Fig.9d) and with the weft threads of the lower layer (Fig.9d). Then we transfer the weaves of Fig. 9a and Fig. 9b to the filling pattern, and set the lifts of the upper and presser warp yarns (O) when forming the bottom layer of the fabric.

Two-ply weaves with pressed weft binding

The special feature is that each presser weft is tied to one of the warp threads of the upper and lower layers of the fabric, and three weft systems (upper, lower and presser) and two warp systems are required (upper and lower layers). The ratio between the yarn systems in the fabric can be varied. The weave rapport in a fabric is determined by the product of the smallest multiple of the base weaves of the layers by the sum of the ply ratios. This takes into account the weave pattern of the pressed weft with the upper and lower warp yarns. The weft is double-waisted in the remiz and the upper and lower warp yarns are woven into the reed tooth.

Let's build a filling pattern of a two-layer fabric with pressed weft on the basis of twill 2/2, the ratio of weft yarn systems 1:1:1 and 1:1 of warp yarns. The weft pattern: $R_y = (1 + 1 + 1) \text{-Re} = (1 + 1 + 1) \text{-}4 = 12$.

Ramp of the fabric on the base: Ro = (1 + 1) -Rs = (1 + 1) -4 = 8. The number of heddles in the dressing **k** = 8, the purl in the heddle is double-wired and two warp threads of the upper and lower layers are picked into the reed tooth. Figures Yua and 106 show the weaves of the outer side of the upper layer and the inner side of the lower layer. In the cross section of the fabric (Fig. Jv), we determine the possible ligatures of the presser weft with the upper warp and with the lower warp.

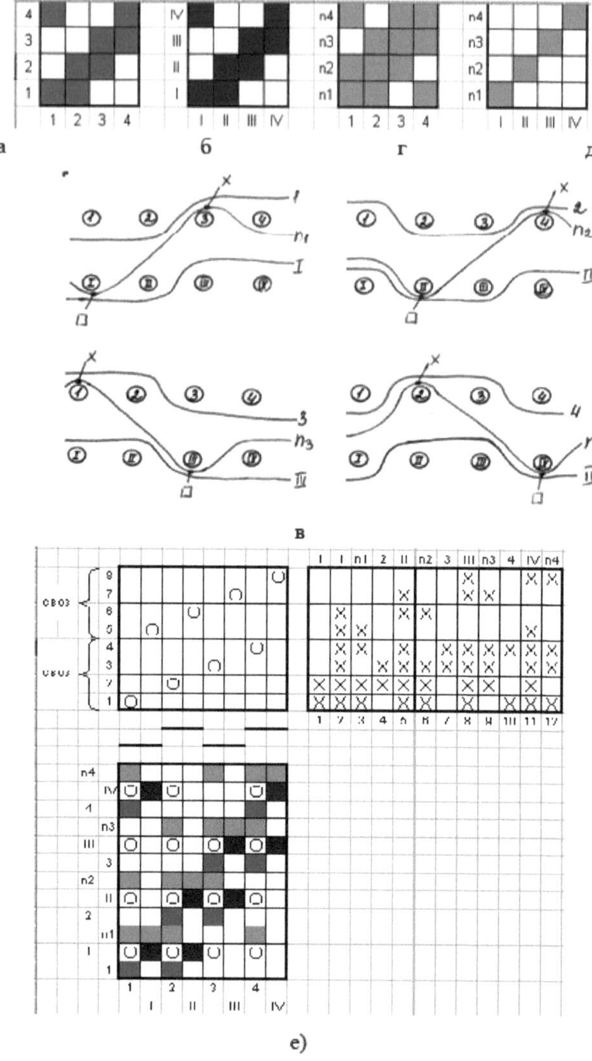

Figure 10. Weave of a two-layer fabric by tying the layers with a pressed weft: a - weave on the outer side of the upper fabric layer; b - weave on the outer side of the lower fabric layer; c - cross section of the fabric and places of dressing of layers by pressing warp with upper weft and lower weft; d - programme of dressing of layers in the fabric by pressing weft threads with upper warp threads; e - programme of dressing of fabric layers by pressing weft threads with lower warp threads; f - filling pattern of two-layer fabric by dressing of layers by pressing weft.

There are four possible ways of tying the presser weft to the warp yarns for the

first fabric cut.
The first is to tie the presser weft to the third upper warp thread and to the first lower warp thread.
The second is the tying of the presser weft to the fourth upper warp thread and to the second lower warp thread.
The third is the tying of the presser weft to the fourth upper warp thread and to the first lower warp thread.
The fourth is the tying of the presser weft to the third upper warp thread and to the second lower warp thread.
The cross section shows a variant for the first filling yarn, where the presser weft is tied to the third upper warp and to the first lower warp. Let us denote by (X) the places where the presser weft is tied to the upper warp layer, i.e. in this place the upper warp yarns are under the presser weft. Let us denote by (□) the places where the presser weft is tied to the lower warp yarns, i.e. at this point the warp yarns of the lower layer are above the presser weft. On the basis of the cross section of the fabric and the places where the layers are tied by the presser weft with the upper weft and the lower weft (Fig. 10c), we build a programme for tying the presser weft with the upper warp yarns (Fig. 10d) and with the lower warp yarns (Fig. 10d). Then we transfer Fig. 10e, Fig. 106, Fig. 10g and Fig. 10d to Fig. 10e in the filling pattern of the double-layer fabric by tying the layers with the presser weft and put the lifts (O) of the upper warp yarns in the lower wefts and presser wefts.
Sew one upper and lower warp thread into the reed tooth. We adopt a two-wire sieving in the remise, i.e. all the upper warp threads are sieved into the first arch and all the lower warp threads are sieved into the second arch.

Conclusion

Two-ply fabrics are produced on weaving machines equipped with eccentric and dobby shedding mechanisms, multicolour or multicloth devices. When weaving fabrics with identical wefts in the layers, single-colour or single-row looms are used. Single or double warp threading machines can be used depending on the warp yarns in the layers. When producing fabrics with a pressed warp, only double warp threading is necessary, as the pressed warp has a higher yarn count. The warp threads are usually picked into the ream, with the warp threads of one layer being picked into one vault and the warp threads of the other layer being picked into the other vault. In the reed tooth, the warp threads of each layer are sieved. two-layer fabrics are used as clothing fabrics (drapes) and technical fabrics (kirza, technical tapes, etc.).

CHAPTER 6

6. MULTILAYER WEAVES

Widely used in the production of technical fabrics, technical felts, drive belts, filter fabrics, kirz, etc. Multilayer fabrics have high resistance to tearing, piercing, friction. Special weaving machines are used to produce such fabrics. Multilayer fabrics have at least three warp and three weft systems. Each warp system intertwined with the weft forms an independent layer of fabric and the individual layers joining and tying together form a single whole fabric. There are the following tying methods:

1.Dressing of layers from top to bottom, i.e. from top to bottom (Fig.1).

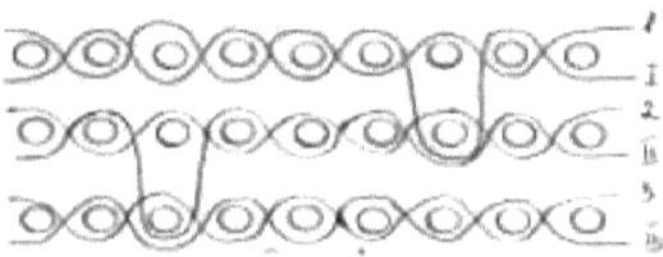

Figure 1. Dressing of fabric layers from top to bottom.

2.Dressing of layers from the bottom layer to the top layer, i.e. from bottom to top (Fig.2).

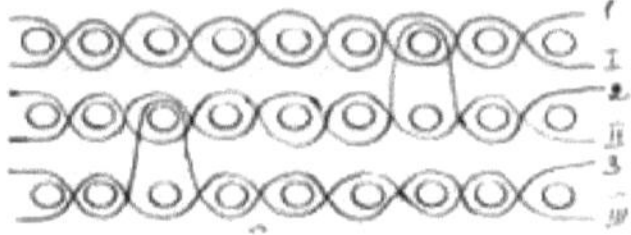

Figure 2. Dressing of fabric layers from bottom to top.

H.Dressing of layers from top to bottom layer and from bottom to top layer, i.e. combined (Fig.H).

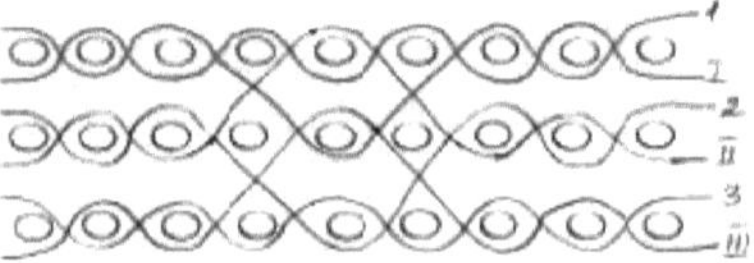

Fig.Z. Combined dressing of tissue layers.

4.Dressing of layers with pressing threads (Fig.4).

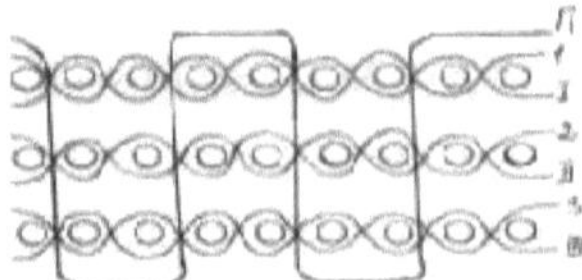

Figure *4*. Layer dressing with clamping threads.

When pressed warp layers are joined, the warp yarns have a high workmanship, so the pressed warp is wound on a separate warp warp, which complicates filling and machine maintenance. The warp length in multilayer fabrics with pressed warp yarns is determined by the formula Ro = 2is + n_{pr}, where: Is - number of layers; n_{pr} - number of pressing threads in the rapport.

The length spread on the base of a multilayer fabric with layers tied to each other is determined: Ro = 2-nc. Rape in the weft of a multilayer fabric: $R_y = {}_{R(yc)}$- n_c where: Rye- the weft pattern in one ply.

The graphical representation of a multilayer weave is mainly made on the basis of a longitudinal section of the fabric, in which the number of plies, the weave in the plies, the nature of the ply binding, the warp and weft ratio are determined. The number of plies is determined by the number of rows of weft yarns. The filling pattern is based on a longitudinal section of the fabric. The warp and weft threads of all layers are numbered with Arabic numerals. The warp yarns are picked through the weft threads in a row, and the number of threads equal to the warp rapport is picked through the reed tooth. The number of loops (k) is equal to the warp pattern **($R_{o)}$, k = R_o**

Number of remisks on production of multilayer fabric with pressed warp: k = Ro+ ${}_{n(n)}$, where: n_n- number of pressed warp threads in the fabric rapport.

Multilayer weaves with combined ply binding

Draw the weave pattern of a three-layer fabric. Weave of each layer is plain weave. The binding of layers is combined. The ratio of the number of yarns of separate layers on the warp on the warp and on the weft is 1:1:1.

Base rapport: Ro= ${}_{2\text{-}n(c)}$= 2-3 = 6. Weft pattern: R_y= ${}_{R(yc)}$-n_c= 4-3 = 12.

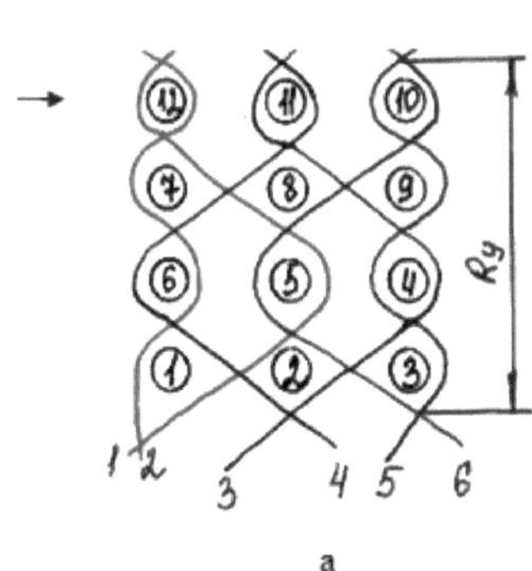

a

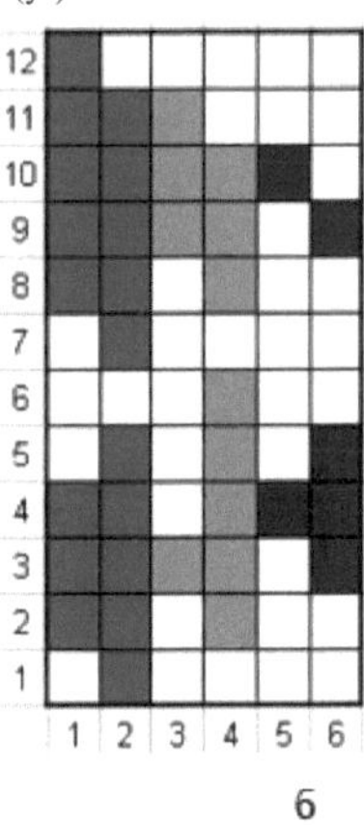

б

Figure 5. Weave of a three-layer fabric by combined ply binding
a - longitudinal section of a three-layer fabric; b - filling pattern of a three-layer fabric by combined dressing of layers.

From the longitudinal section (Fig. 5a), it follows that when laying the first weft, the second warp thread is picked up, and when laying the second weft, the first, second and fourth warp threads are picked up, and when laying the third weft, the first, second, third, fourth and sixth warp threads are picked up, etc. The above is transferred to the threading pattern (Fig. 5b) of a three-ply fabric.

The sashing in the remiz is row by row for six remiz, the whole rapport of warp yarns, i.e. six warp yarns, are sashing in the reed's tooth.

Multilayer weaves with pressure backing

Let's build the weaves of a three-layer fabric. The weave of each layer is plain weave. The ratio of the number of threads of the layers in the warp and weft is 1:1:1.

Rapport on the base: Ro = 2-Ns + n_{pr} = 2-3 + 2 = 8.

Duck rapport $R_y = R_{yc} \cdot n_c = 2 \cdot 3 = 6$.

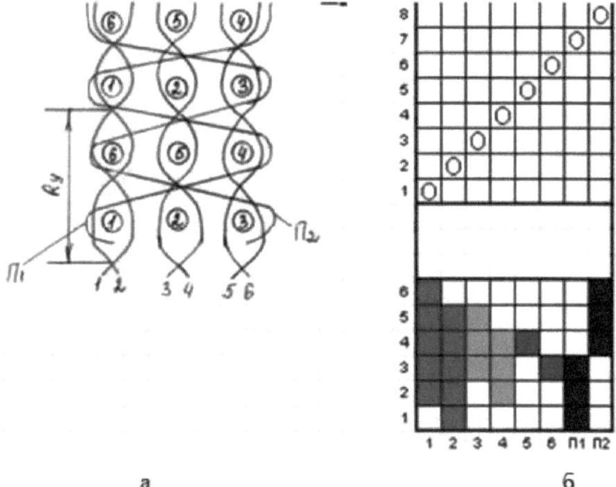

Fig. 6. Weave of a three-layer fabric with pressure-base binding: a - longitudinal section of a three-layer fabric; b - filling pattern of a three-layer fabric with pressure-base binding.

From the longitudinal section (Fig. 6a) it follows that the second warp and the first presser base (P1) must be lifted during the first weft weaving, the first, second, fourth and first presser bases must be lifted during the second weft weaving, and the first, second, third, fourth, sixth and first presser bases must be lifted during the third weft weaving, etc. During these periods of weft laying, the second presser base (W) is in the lower position. The above mentioned reasoning is transferred to Fig. 66 and the filling pattern of the three-ply pressed warp is drawn up.

The number of heddles in the dressing is eight, the warp purfling is row, the purfling in the reed tooth is eight warp threads.

Conclusion

Multilayer fabrics are produced from yarns with high linear densities. For the production of these fabrics, heavy weaving machines with single, double and triple warp threading are used depending on the number of layers, warp and weft densities in the layers, warp yarn processing and the method of tying the layers. Usually, the warp threads are picked in the remiz row, and the warp threads are picked in the reed tooth of the whole warp rapport on the warp.

CHAPTER 7

7. PILE WEAVES

Pile fabrics are fabrics that have a pile cover on the surface made of densely standing thread tips. Not to be confused with tufted fabrics such as baik, flannel, boomzea.

According to the nature of the pile surface formation, fabrics are subdivided:
1.Topstitch pile, in which the pile is formed from the tips of the weft threads.
2.Basic pile, in which the pile is formed from the tips of the main threads.

Fibre wool fabrics

One warp system and two weft systems are required to form weft-wool fabrics.
In the process of fabric formation, weft skeins are subdivided into ground skeins, which form the fabric ground with the base and pile skeins, from which the pile is formed when finishing the fabric (Fig.1). Usually as a basic ground weave the plain weave, twill weave are used, and as a basic pile weave the satin weave is used. The most widespread are duck-wool fabrics such as corduroy-cord (pnc.la), corduroy-rubchik, semibarchat (Fig. 16). Semi-barchat is characterised by a uniform arrangement of pile on the surface of the fabric, and corduroy-rib differs from corduroy-cord by the rapport on the warp and weft. Weft-wool fabrics are used for sewing clothes, shoes, etc.

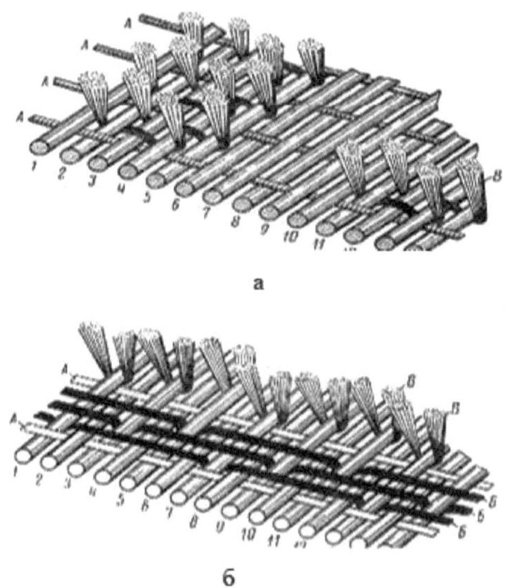

Fig.1. Schematic of the structure of weft-wool fabric: a - corduroy cord; b - semi-velvet; A - ground (root) weft; B - uncut pile weft; C - cut pile weft; 1,2,3, etc. - warp yarns.

The base rapport Ro of a weft-wool fabric is defined as the smallest multiple of the basic weave base rapports for the ground (root) Ror and pile R_{OB}, i.e. when Ro = 6 Ror= R_{OB} = 2 or 3.

When selecting weaves based on the base of the basic ground (root) weft Ror, it is necessary that it be equal to or a multiple of the base pile weft rapport Ro_B, i.e. with Ro_B = 6 Ror = 2 or 3.

The base rapport of the basic pile weft Ro_B is equal to the sum of the main threads $No_{(B)}$, which it overlaps in forming the pile deck and the main threads Noz, by which it is secured $R_{(0B)} = n_{(OB)} + n_{03}$

The weft rapport R_y of a weft-wool fabric is defined as the product of the weft rapport of the ground weave $R_{(yr)}$ and the sum of the ratio of the number of threads of the ground weft Nu_G and the pile weft Pu_C. Ry= Ry_r (Pu_G + $Pu_{(B)}$)), where: Nu_G is the number of ground weft yarn ratios; n_{uv} is the number of pile weft yarn ratios.

The ratio between the number of yarns of the ground and pile wefts can be from 1:2 to 1:6 or more.

Let's build the filling pattern of the weave of corduroy-cord fabric.

The filling pattern and cross section of the fabric is shown in Fig.2, the fabric has a pile in the form of stripes on the surface.

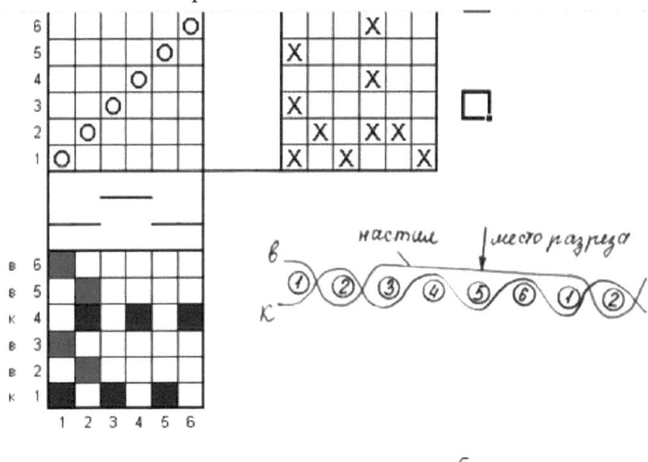

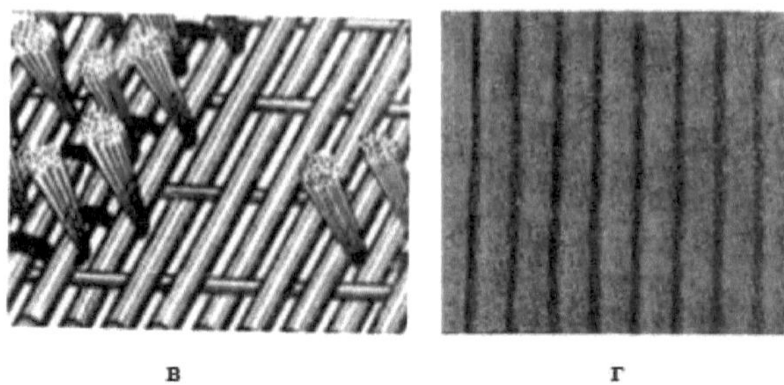

Fig.2: a - full filling pattern of corduroy-cord fabric; b - cross section of corduroy-cord fabric; c - scheme of corduroy-cord fabric structure; d - view of finished corduroy-cord fabric.

The weave of the ground is plain weave. Length of weft weft $n_{0B} = 5$ warp yarns, number of basic yarns secured by pile wefts $n_{03} = 1$, ratio of pile wefts ($n_{(uv)}$)) to ground wefts ($n_{(ug)}$)) 2:1. We determine the rapport on the base of the basic pile weft:

$$R_{OB} = n_{OB} + n_{O3} = 5 + 1 = 6.$$

RAMP on the base of the basic ground weft: R_{O1}2.

Weft rapport of the basic ground weft: $R_{(yr)}$) = 2.

Rapproportion of the quilting fabric on the base: $R_O = R_{O_B} = 6$.

Ramp of weft fleece: $R_y = R(yr) (n_{(ug)} + n_{uv}) = 2 (1 + 2) = 6$.

Figure 3 shows the complete tuck pattern and cross section of a semi-barchatt fabric with a continuous pile cover and a 4:1 ratio of pile to ground weft.

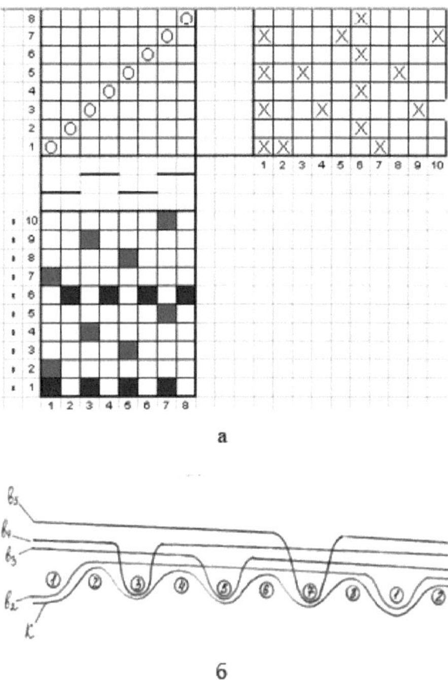

Fig.3. Weave of duck-wool fabric half-barchat: a - full dressing pattern of fabric half-barchat; b - cross section of fabric half-barchat.

The weave of the ground plain, pile weave is constructed on the basis of irregular sateen octamite. Rapport of weave of duck-wool fabric semi-velvet on the base $Ro = {}_{R(OB)} = 8$. Rapport of weave of duck-wool fabric semi-velvet on weft: $R_y = {}_{R(yr)}) (n_{(yr)} + {}_{n(yB)}) = 2 (1+4) = 10$.

The number of heddles in the dressing is eight. The sashing in the heddles is rowwise, the sashing in the reed tooth is two warp yarns each. All weft wool fabrics are produced from cotton yarn of low linear density with very high weft density (1000 threads per Y cm. and higher).

Basic linen fabrics

In warp yarns, the pile is created from the warp yarns directly during the weaving process. The pile yarns are cut by a special cutting mechanism. Varieties of basic pile fabrics are: velvet, which has the lowest (up to 2 mm.) and the most dense vertically standing pile, which covers the fabric ground from the front side; plush, which has a higher (2-4 times higher than velvet), but less dense pile, which due to its high height is arranged obliquely, which contributes to covering the fabric ground from the front side; artificial fur with a pile height of more than 10 mm.

According to the method of pile formation, basic pile fabrics can be:

- tufted fabrics in which the pile surface is formed from stretched pile loops firmly anchored in the ground;
- slit, fabrics in which the pile surface is formed from thread tips protruding from the ground;
- stretch-cut, fabrics in which the pile surface is formed from loops and protruding thread tips.

According to the method of arrangement of the pile cover, basic pile fabrics can be:

- smooth pile, fabrics with a continuous smooth pile surface;
- fine-patterned, fabrics with a combination of a pile smooth surface with a small pattern;
- coarse-patterned, with a combination of smooth and pile weave with a large pattern spread.

In terms of fibre composition, basic wool fabrics can be made of wool, natural silk, cotton, linen, viscose, lavsan and other fibres.

According to the method of loom formation, warp fabrics can be single-weave and double-weave.

Single web fabrics are produced on rod machines. Depending on the shape of used rods, fabrics with drawn pile (round-shaped rods), with cut pile (oval-shaped rods with a blade at the ends) and drawn and cut pile (round and rods) are produced. The initial data for construction of the filling pattern is a longitudinal section of the fabric, which is used to determine the type of ground weave, the ratio of yarn systems and the method of fixing the pile in the fabric.

The ground weave of warp pile fabric is usually plain weave, main rep 2/2, main half-rep 2/2, weft 2/2. The ratio of the number of threads between the warp and pile warp can be 1:1:1, 2:1:1:1. The pile warp can be fixed in each fabric with one, two, three or more wefts.

Fig. 4a shows a longitudinal section of a pile fabric formed by the rod method. The ground is formed by a 2/1 half-repeat weave (yarns 1,2,3,4, Fig. 46) and the pile yarns by plain weave (Fig. Zv yarns Bi and $B_{2)}$. During weft shedding, the pile yarns are earned together with the warp yarns into the fabric ground. After three weft shedding operations, the pile yarns rise and then fall down during the next shedding operation and are placed on the loop rod P1. Together with the fourth threading, the rod is nailed to the fabric edge. In this case, a transverse row of loops is formed on each bar across the entire width of the filling, which are cut with a blade when the bar is pulled out. If the rods are not cut with a blade, pile loops remain in the fabric. The number of reams in the dressing is six (Fig. 4g), the purl in the ream is cumulative, the purl in the reed tooth is three warp threads each.

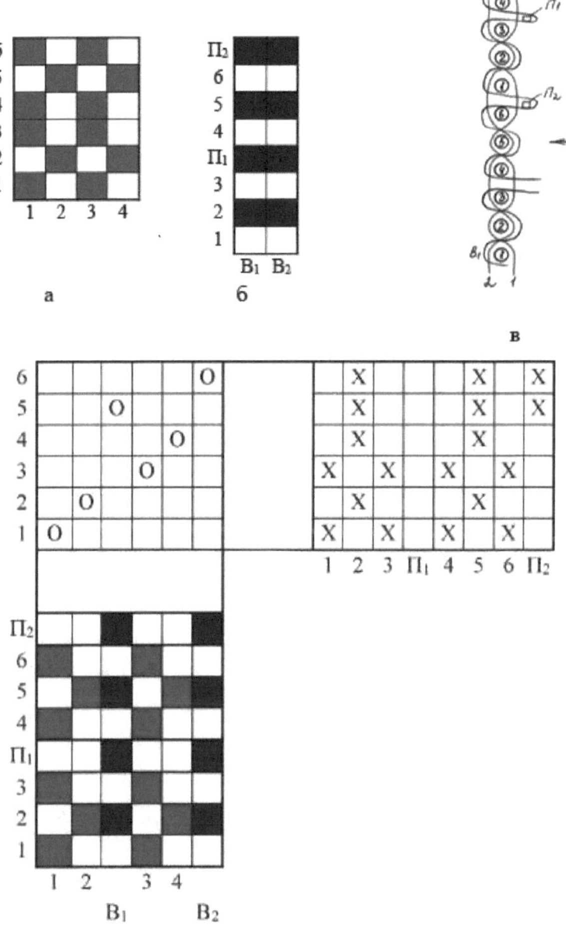

Fig.4. Single web warp fabric: a - longitudinal section of single web warp fabric; b - ground weave of fabric; c - pile weave of fabric; d - full filling pattern of single web warp fabric.

The two-weave twin-sewing method of warp weaving is characterised by the presence of two weft yarns (upper and lower weft weft) and three main yarns (upper weft, lower weft and pile) in the system.

The essence of the production of these fabrics is that the wefts of the upper and lower webs are simultaneously laid in two sheds (upper and lower). The laid wefts form two webs simultaneously with the root threads, which are connected by the pile warp threads passing from one web to the other web. When the fabric is diverted, the pile warp threads between the webs are cut, resulting in two webs of pile fabric.

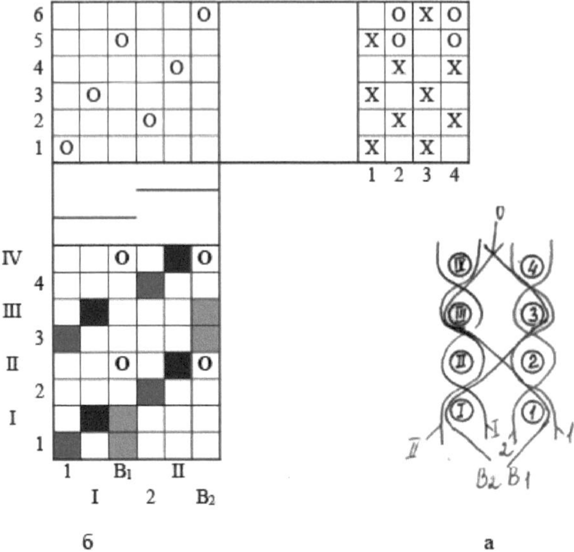

Fig.5.Two-weave worsted fabric: a - longitudinal section of two-weave worsted fabric; b - complete filling pattern of two-weave worsted fabric.

Fig. 5a shows a longitudinal section of a two-weave, two-zewn pile fabric. As can be seen, warp yarns 1 and 2 of the upper web and warp yarns I and II of the lower web are interwoven respectively with the upper weft yarns in a plain weave, as shown in Fig. 56. The pile yarns Bi and B_2 are interwoven with the ground of the upper and lower fabrics (with wefts 1 and 3) in a basic rep 2/2 weave. And at laying of wefts 2, II, 4 and IV the pile warps come to the middle position, i.e. they are located between the webs, this position is marked with (0) in the filling figure Fig. 4b. The design of the machine ensures this middle position of the pile warps. The number of threads in the threading is six, three threads in each thread, three threads in each thread, and three threads in each tooth of the reed. The ratio between the warp systems (upper warp, lower warp and pile) is 1:1:1.3 The pile warp is attached by one stitch in each rapport. The weave pattern in the fabric is $R_o = 6$ in the warp and $R_y = 8$ in the weft.

The warp yarns are picked through the reed tooth with a number of yarns equal or multiple to the sum of the ratio of the number of yarns in the warp.

Conclusion

Fill-wool weaves are designed for clothing and furniture decorative fabrics and are produced on conventional weaving machines equipped with powerful shedding mechanisms and warp release and tensioning mechanisms. Row sashing, warp sashing in the reed tooth of 2-3 threads are used. Basic pile weaves are intended for clothing, drapery and other fabrics, and are produced on special looms equipped with two-way (multi-way) threading mechanisms of warp release and tension for warp and pile threads, mechanisms of insertion and extraction of bars (in the rod method of fabric formation), a device for cutting pile threads between webs at fabric withdrawal.

8. BUTTONHOLES

Quilted weaves are a type of warp fabrics. They are divided into one-sided (loops only on the front side of the fabric) and two-sided (loops on both sides of the fabric). Quilting (terry) fabrics are produced from three yarn systems - warp and quill warp and weft yarns. Root warp thread is interwoven with weft (basic rep 2/2 and 3/1, etc.) to create a ground basis of the fabric, in which the threads of the quill warp are fixed by basic rep 2/2, basic half-reps 2/1 and 3/1, etc. The main warp thread is interwoven with weft. The ratio between the number of warp and loop warp threads is 1:1,2:1,1:2.

The following process conditions are necessary for the formation of loops on the fabric surface:

1 .Different warp (maximum) and buttonhole (minimum) tension.

2 .Placed weft yarns within the weft rapport shall have two incomplete fetches (Fig. 1) for each revolution of the machine main shaft at a distance equal to twice the pile height **2h.**

3 .Full surf shall be made after laying the last utochina of each rapport (in Fig. 1, third utochina-3).

4 .At full surf, all the wefts of the rapport (in Fig. 1, the first-1, second-2 and third-3 wefts) should slide along the stretched warp threads 1 and 2 at a distance of **2h,** entraining the weakly stretched loop threads *Bi* and *B 2,* forming loops the height of which is equal to h.

5 If the same positioning of the buttonhole base in relation to the weft (Fig. 1, third point-3) of the full surf and the following first weft of the incomplete surf group (Fig. 1, first point-1), loops are formed on both sides of the fabric. If the buttonhole base overlaps both of these two wefts, a loop is formed on the upper side of the fabric, and if the buttonhole base is lowered under these two wefts, a loop is formed on the lower side of the fabric.

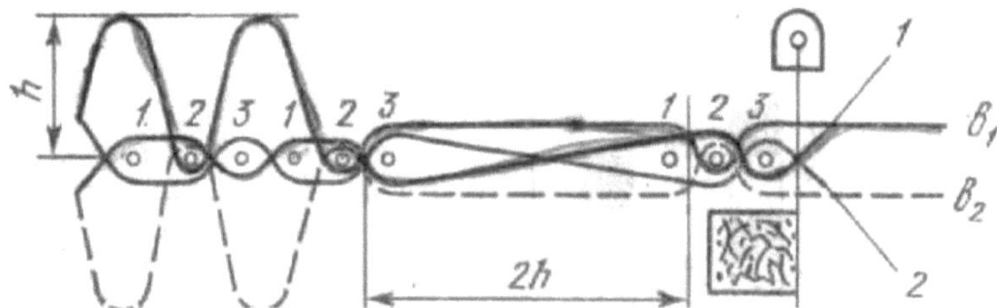

Figure 1. Schematic of loop formation on the fabric surface.

Quill (terry) fabrics are produced on special machines with variable batana stroke, devices that reduce the tension of the quill warp and increase the tension of the warp threads at full drowning surf.

Let's build the filling pattern of a double-faced terry cloth. The weave of ground (root) and loop warp with weft threads is semi-repeat main 2/1, the ratio of the number of loop and terry threads is 1:1. Rapport of the fabric on the weft $R_y = 3$, rapport of the fabric on the base Ro = 4. For better fixation of the loops in the fabric, the loop warp weave is shifted in relation to the warp weave by one thread at odd weft rapport and by two threads at even weft rapport. Since in our example the weft rapport has an odd value, we will shift the loop warp weave in relation to the warp weave by one thread (Fig. 2). The warp threads are picked into the remise in a cumulative manner, with two threads in the reed tooth - one warp and one buttonhole warp.

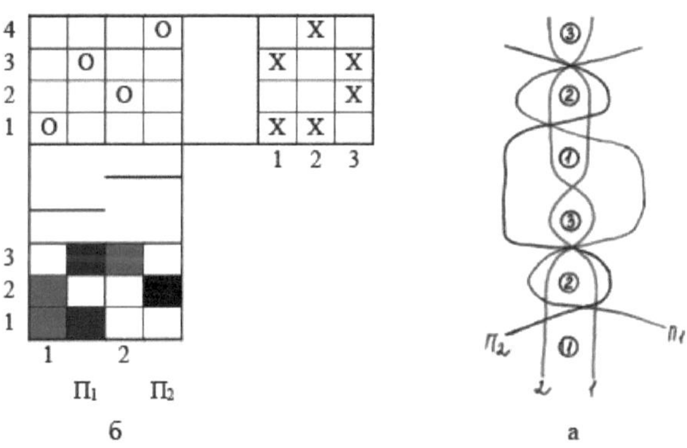

Figure 2. Two-faced terry cloth weave: a - longitudinal section of double-faced terry cloth; b - full tucking pattern of double-faced terry cloth.

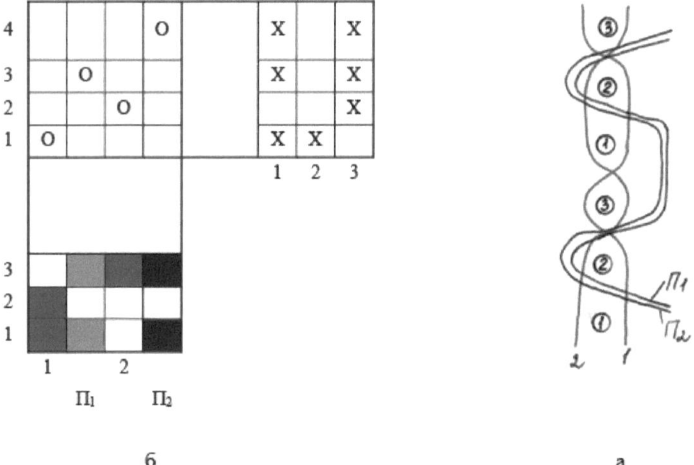

Fig.3. Weave of double-sided terry cloth: a - longitudinal section of double-sided terry cloth; b - complete filling pattern of double-sided terry cloth.

Fig.Z shows a longitudinal section and a complete filling pattern of a double-faced terry fabric with one-sided buttonhole, based on the parameters of the previous double-faced terry fabric. By combining smooth and buttonhole patterns on the upper surface of the fabric, a variety of patterns can be obtained.

Conclusion

Terry fabrics are used for making sheets, bathrobes, towels, mats, etc. Terry fabrics are produced on special weaving machines equipped with double warp threading, with variable surf (two, three incomplete and one full ducking surf), systems providing maximum tension of the warp and minimum tension of the quill warp, devices for the necessary feeding of the quill warp at full ducking surf. the necessary delivery of the warp at full weft surf. For the buttonhole warp, yarns of lower twist than for the warp are used. Picking in the remiz is summary. A number of yarns equal or multiple to the sum of the ratios between the warp and buttonhole warp are picked into the reed tooth.

CHAPTER 9

9. PIQUÉ WEAVE

There is a simple and a complex piqué weave. In a plain piqué weave, three yarn systems are involved - the face warp, the back warp (quilting) and the face weft. Complex piqué differs by the fact that in addition to the facial weft the laying weft is used. A characteristic feature of piqué fabrics is that on the front side of the fabric a relief and convex pattern in the form of transverse scars, rhombuses, cells, the contour (borders) of which is drawn into the fabric, which resembles the pattern of a quilted blanket, dressing gown, is clearly distinguished. Therefore, the name piqué comes from the French, which means quilted, stitched.

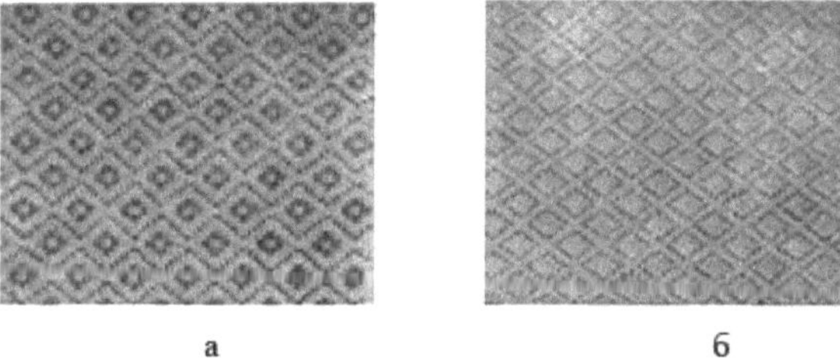

Fig.1.View of finished piqué fabric: a - simple piqué; b - complex piqué.

The face warp and the face weft are interwoven to form a face fabric, usually of plain weave. The root (quilting) warp overlaps the face weft along the outlined contour with short overlaps, located mainly on the underside of the fabric in the form of long and sparse main decks. The peculiarity of the piqué fabric production technology is a normal tension of the front warp in the dressing and a very high tension of the main warp, which causes tightening of weft threads to the underside of the fabric. The ratio of front to warp densities is 2:1, and the ratio of weft to lining is 2:1 (for complex piqué).

The warp R(y) is equal to the product of the smallest common multiple of the warp R(y) of the pattern motif and plain weave multiplied by the sum of the thread ratios in the system. The weft rapport R_y is equal to the smallest common multiple of the pattern motif and plain weave (for simple piqué), and is equal to the sum of the face $R_{(yjI)}$and lining $R_{(yn)}$) weft rapports (for complex piqué). $R_y = R(yjI) + R(yn)$). The face weft rapport R_{yjI} is defined as the total multiple of the weft rapports of the pattern motif and plain weave. The lining weft rapport R_{yn} is defined as the front weft rapport divided by the ratio of the front and lining weft rapports. For example, if the ratio of face weft to lining weft is 2:1, then the lining weft rapport will be $R_{yn} = R(yn)/2$.

A simple dive

To construct a filling pattern for piqué fabric, you must select a pattern motif and draw it on the canvass paper.

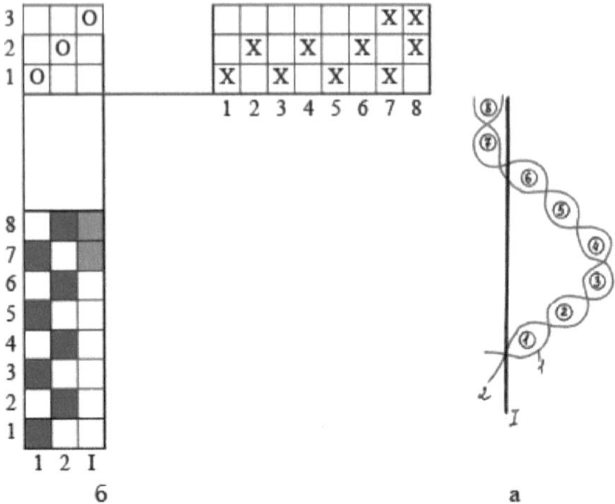

Fig. 1. Weave of plain piqué-rubber fabric: a - longitudinal section of plain piqué-rubber fabric; b - complete filling pattern of plain piqué-rubber.

The simplest example is a piqué fabric - a welt, i.e. a contour of short overlaps of the root warp of the face weft in a straight line across the width of the fabric and from the desired size (width) of the welt under which the root warp is located.

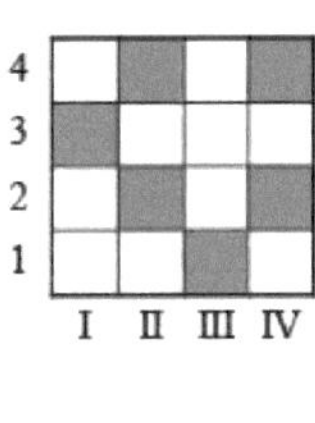

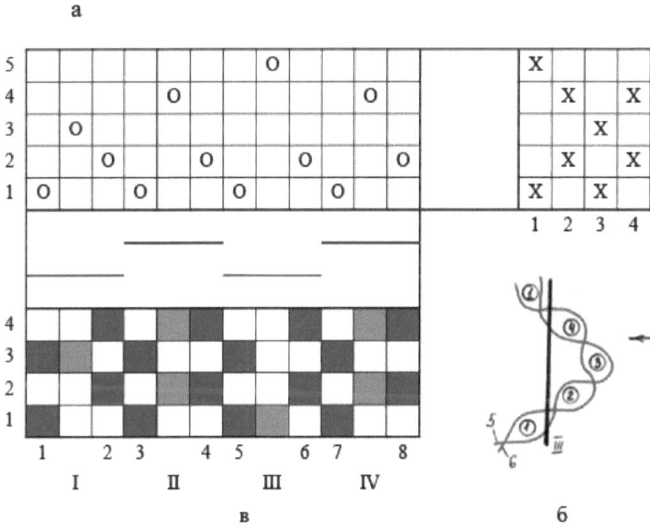

Fig. 2. Weave of plain piqué fabric: a - pattern motif of plain piqué fabric; b - longitudinal section of plain piqué fabric; c - full filling pattern of plain piqué.

When the ratio of face warp to root warp is 2:1. Simple piqué rapport on the warp Ro = 3. Rape of plain piqué on weft (depending on the desired pattern) we take Ry = 8. Fig.1 shows the complete filling pattern of a simple piqué and a longitudinal section of a simple piqué - . The threads are picked in the remiz row by row, and in the reed tooth by three threads. Fig.2a shows the motif of the plain piquet pattern, by which we determine the pattern's rapport. Roy= $_{R(y)}$ = 4. Rape of the weave on the warp at the ratio of the face warp to the root warp 2:1. Ro= $_{R(oy)}$) (2+ 1) = 4(2+ 1) = 12.

In the weave of the fabric (Fig. 2c) we build a plain weave for the front warp, and then transfer from Fig. 2a the pattern motif for the root warp. Thread picking in the reed tooth is three threads each, and thread picking in the remise according to the pattern for five remises. Fig. 26 shows a longitudinal section of the piqué fabric for the third warp thread.

A difficult peak

For greater relief (convexity) of the pattern on the fabric, an additional lining weft (somewhere on the order of an order of threads thicker) is introduced.

Complex piqué is worked on a multicolour machine. Let's show the construction of a complex piqué - welt of the same width (Fig.1), with weft alternation 4:1. Rapport on the base Ro = 3. Rapport on the weft $R_y = R_{(yjI)} + R_{(yn)}) = 8 + 2 = 10$.
In the shed when laying the lining weft, all face warps are raised and all root warps are lowered, so we put the rises (O) in Fig.Z. The lining wefts are inserted before the face weft 7. The quilting pattern in the piqué can be arranged according to an arbitrary motif, so when constructing a complex piqué according to the motif, it should be borne in mind that the thread of the underlying warp overlapping the front weft, the weft according to the motif also overlaps the next lining weft.

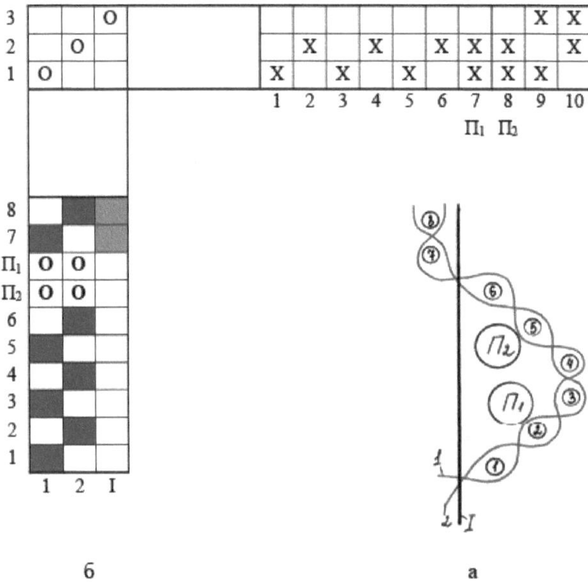

Fig. 3. Weave of the fabric complex piqué-rubber: a - longitudinal section of the fabric complex piqué-rubber; b - full dressing pattern complex piqué-rubber.
Fig. 4a shows a diamond twill pattern motif. In Fig. 46 is a tuck pattern of a complex piqué fabric, and Fig. 4c is a longitudinal section of the fabric. The ratio of face weft to lining weft is 2:1. The warp to backing ratio is 2:1 face to backing ratio.

$$Ro= Roy\ (2+ 1) = 4(2+1)= 12.$$

Duck rapport:

$$Ry\ Ryn\ Ryn = + = \ Ryn + Ryn/ 2 = 4 + 4 / 2 = 6.$$

The lining wefts are laid after the second 2 and fourth 4 face wefts. In both cases, three warp threads are threaded into the reed tooth.

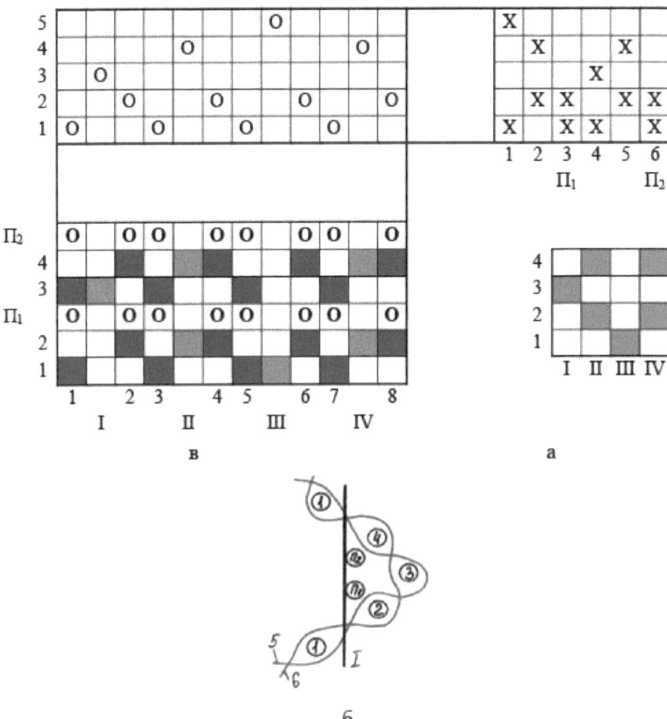

Fig. 4. Complex piqué weave: a - pattern motif of complex piqué fabric; b - longitudinal section of complex piqué fabric; c - complete filling pattern of complex piqué.

Conclusion

Piqué fabrics are produced from yarns of low linear density and are used for making summer suits, dresses, blankets, etc. To insulate fabrics, the wrong side is subjected to fleece. Piqué fabrics are produced on machines with two-napped threading and multicolour feeding of the weft (for complex piqué). As shedding mechanisms, dobbies or jacquard machines are used.

CHAPTER 10

10. OPENWORK AND EMBROIDERY WEAVES

Openwork fabrics

Fabrics that have an openwork effect on the surface, obtained by twisting the warp threads of one system with the threads of another system are called leno or openwork fabrics (Fig. 1).

Fig.1. Openwork weave fabric.

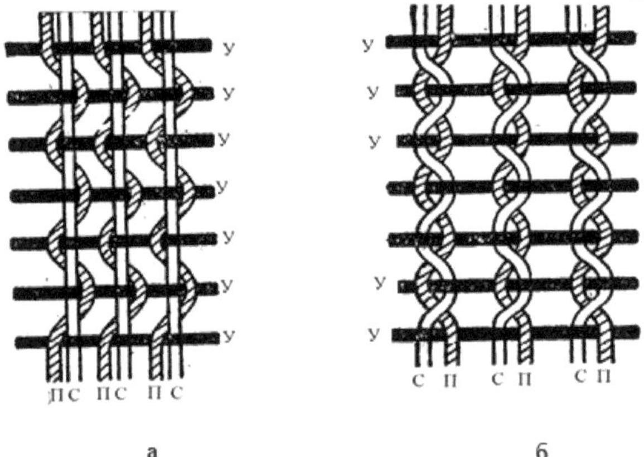

a 6

Fig.2. Ligature weave of openwork fabric, a - before removing the fabric from the loom, b - after removing the fabric from the loom, where P - leno thread, C - stock thread, U - weft.

Openwork effects can be arranged on the fabric across the entire width of the fabric and as individual stripes, squares, checkers, etc. The peculiarity is is that during the weaving process one warp system (leno) is wound on the right and left side of the warp system (maiden). The stock yarns have a high filling tension and the leno yarns have a minimum tension, so the looms have a double filling.

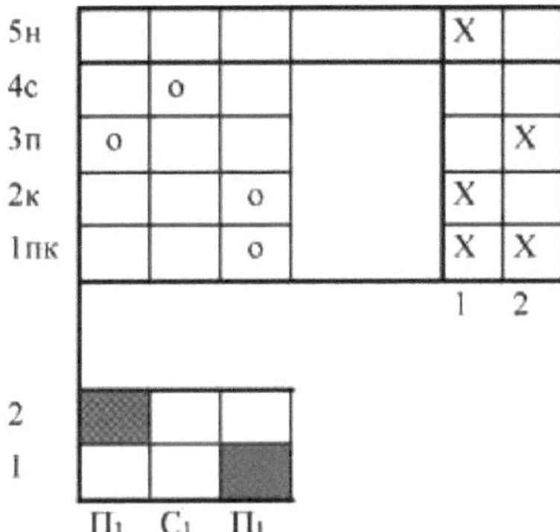

Fig.Z.Full filling pattern of leno weave of simple openwork fabric, where: P - leno thread, S - stockinette thread, 1pk - leno half wing, 2k - leno remizka wing, Zp - leno remizka, 4c - stockinette remizka, 5n - compensating remizka.

The wrapping of the stock yarns with leno yarns is carried out with special leno devices. The following types of leno devices are available.

The first one is shown in Fig. 4, where the loops 1, 2 and 3 of the weft frames are connected to the eye G, in which the leno thread P is inserted. During the first weft cast-off (first shed), loop 2 is lifted and the leno thread P is lifted together with it, located on the left side of the warp C, while loop 1 is lowered, which is made of yarn of low linear density. the second filling yarn (second shed), loop 1 is lifted and with it the leno yarn P on the right side of the stocking warp C and loop 2 is lowered. Loop 3 is used to move 1 and 2 correctly.

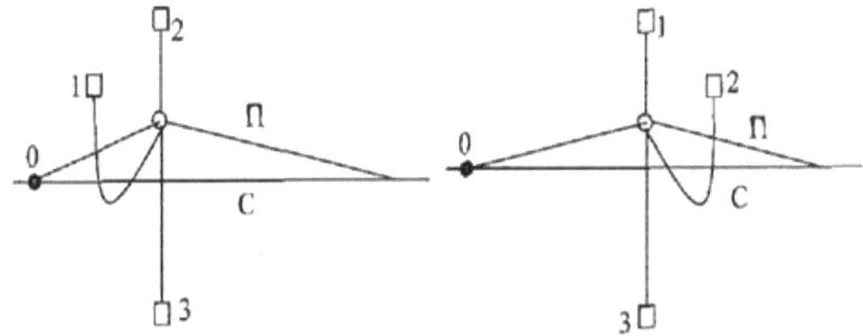

First pharynx Second pharynx

Figure 4. The ligature fixture.

In the second type of devices (Fig. 5), the warp thread is inserted into the eye of the galev of the warp thread loom 1, and the flexible galev (loops) with the eye G carrying the leno thread and covers the warp thread C and is connected to the leno thread looms 2 and 3. At the first weft threading (the first shed), leno 2 moving upwards will position the loop with the eye and together with them the leno thread P on the left side of the stock thread C. The weft thread laid in the shed and nailed to the fabric edge will fix the leno and stocking warp threads in this position.

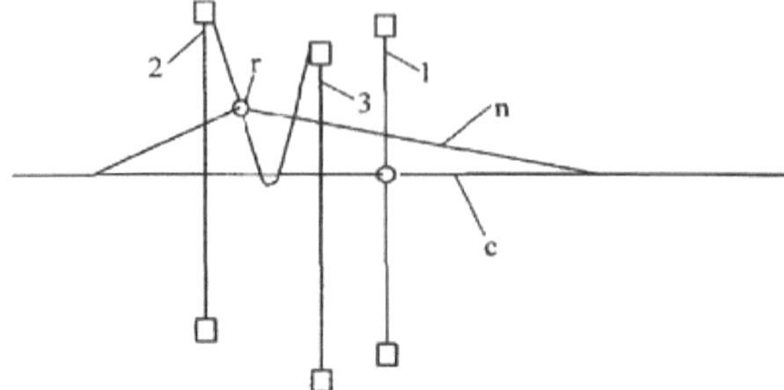

First pharynx

Figure 5. The ligature fixture.

At the second revolution of the main shaft (Fig. 6, change of sheds), the eyelet 3 moving upwards will position the eyelet with loop and as a consequence the leno thread P to the right of the thread C.

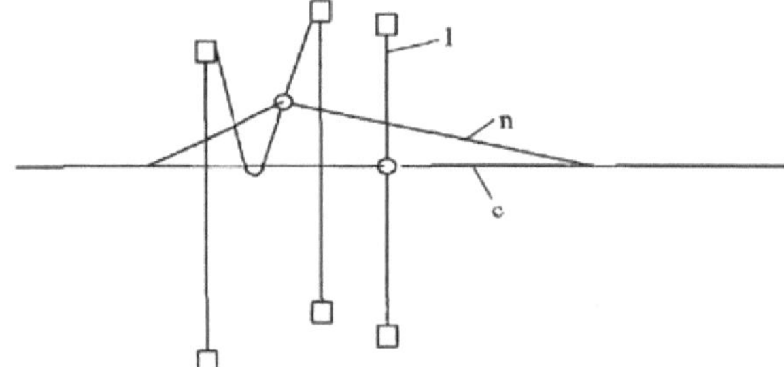

Second pharynx

Figure 6. Ligature fixture

Fig. 7 shows the third type of ligature attachments, where the stock threads C are inserted into the eyes of the ligature 1, and the ligature threads P into the eyes of the ligature 2 and into the ligature pair consisting of a wing 3 and a half-wing 4. Wing 5 is an ordinary remizka in the eye of the halewing which is penetrated by flexible loops (from twisted thread or monofilament) of the half-wing.

At the first yawn (Fig. 7), the leno thread P is positioned to the left of the stock thread C by lifting the leno thread 2 and the half-wing 4, with the wing 3 and the stock thread 1 lowered.

In the second shed (Fig. 8, second utochina), the leno thread P is positioned to the right of the stock thread C, by lifting wing 3 and half-wing 4, with leno thread 2 and stock thread 1 lowered.

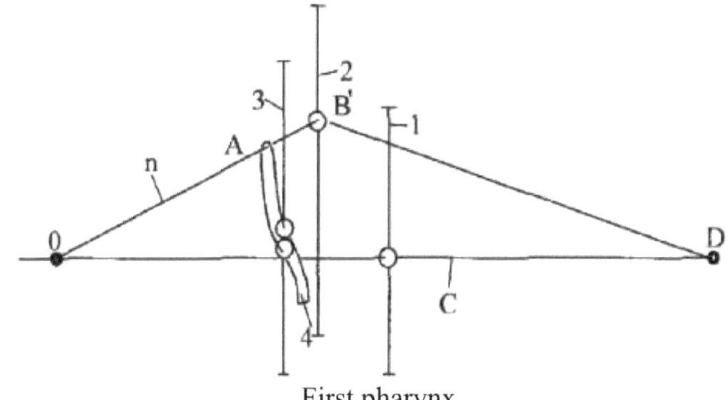

First pharynx

Figure 7. The ligature fixture.

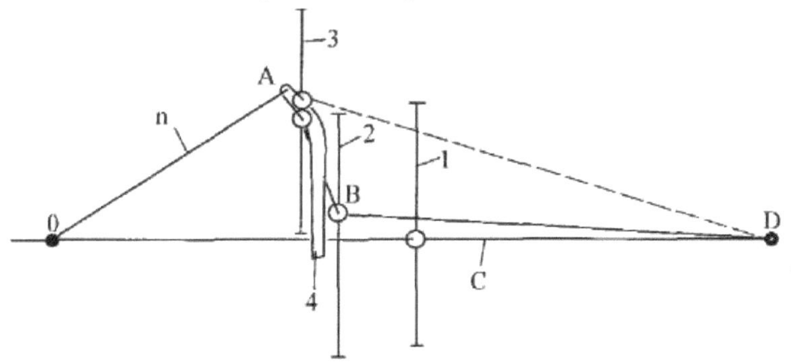

Second pharynx

Figure 8. The ligature fixture.

As can be seen, at each revolution of the main shaft of the machine, the half wing is positioned in the upper position and the rack 1 in the lower position.

The formation of the first shed does not cause high stresses of the leno threads P, since the leno ridge 2 and the half wing 4 are at the top with the geometry of the leno thread dressing OAD.

The formation of the second shed causes a high tension of the leno threads, as the wing 3 and half wing 4 are in the upper position and the leno leno warp 2 is in the lower position, which changes the geometry of the shed filling on the OAWD. To compensate for the length of the leno warp equal to (AB+ BD- AD), the following methods are used:

1 .Installation of an additional forced-oscillating scalo for the leno warp, which loosens the leno warp when a second shed is formed.

2 .Turning the leno with the leno base by a small additional angle when the second shed is formed.

3 .Installation of an additional (compensating) remizki, in the galley of which

the leno threads are penetrated, loosening the leno threads of the warp at the formation of the second shed.

When producing leno fabrics, apart from flexible (threaded) galleys, rigid (metal, plastic) galleys are used.

Fig. 9 shows a metal galewa having an eye in the upper part of the galewa, in which the stock thread C is threaded, the outer surface of the galewa being in contact with the leno thread P.

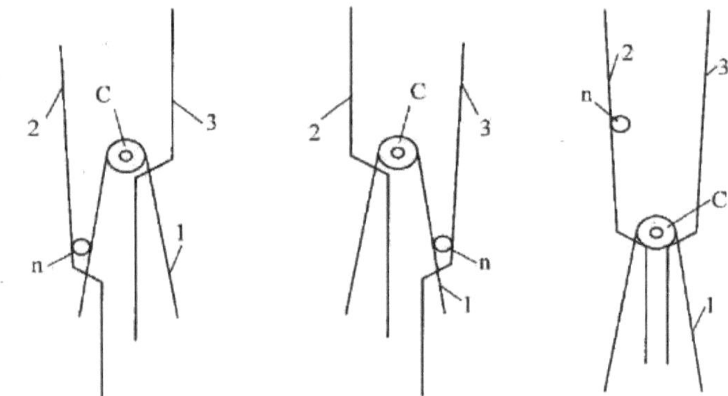

Figure 9. The ligature fixture.

Lowering the left wing 2 causes the leno thread P to be moved downwards and positioned to the left of the stock thread C. Lowering the right wing 3 leads to the downward movement of the leno thread P and its location to the right of the stock thread C. Rigid galleys occupy more space than flexible galleys (threaded galleys) and when lenoing is performed, leno threads may be clamped, which leads to an increase in thread breakage. To produce leno fabrics with no more than 12 threads per 1 cm, special machines are used, in which the stock and grafting galeva (remizki) are replaced by bars with needles, making a lateral movement to form a leno, and then in height to form a shed.

Embroidered fabrics

The construction of such a fabric is very simple: a simple fabric is taken, for example plain weave, and on it in some places some small pattern is placed additionally, executed by warp or weft threads, located next to or close to each other in the form of planking or embroidery. Therefore, fabrics of this type are called embroidered fabrics. The embroidery pattern can be executed with warp or weft threads. Embroidered fabrics are distinguished into warp and weft embroidered fabrics.

Basic embroidery fabrics are produced from at least two warp systems (ground and embroidery) and one weft system (ground). Because of the sharp processing of the warp and embroidery warp, they are wound into separate warps. The embroidery warp threads form a pattern on the surface of the fabric in the form of longitudinal stripes or in the desired pattern with skips (Fig. Yua). The embroidery threads then go to the wrong side of the fabric and can be fixed according to the rule of the underlay warp weave. In jacquard threading, embroidery warp threads are picked into arcade cords, ground warp threads into separate remizki, the number of which depends on the weave of the ground fabric and the type of picking. In the filling of the machine is possible to feed embroidery threads from the bobbins located on the creel at the back of the machine, with a small number of embroidery threads (patterns) on the fabric.

Duck embroidery fabrics have one warp system (ground) and at least two weft systems (warp and embroidery). The production of weft embroidery fabrics requires a special machine with special mechanisms, which allows to introduce different coloured (heterogeneous) embroidery weft threads into the fabric and to adjust the weft density of the fabric (Fig. 10a). This is due to the fact that the number of threads in 10 cm. of embroidery weft is additional to the number of threads in 10 cm. of ground weft, in the formation of ground fabric. If there are sections of fabric without embroidery weft along the length of the fabric, then in these sections the number of threads per 10 cm. of weft will be twice less than in the sections with embroidery weft (at weft alternation 1:1). Therefore, when producing fabric sections without stitched weft yarns, the commodity regulators during this period do not withdraw the fabric from the forming zone or reduce the value of fabric withdrawal by half. This is achieved by switching off or reducing the feed dog travel or reducing the motor speed in the electronic product controller. If the embroidery weft forms a pattern along the entire length of the fabric without interruption, there is no need for variable fabric diversion control.

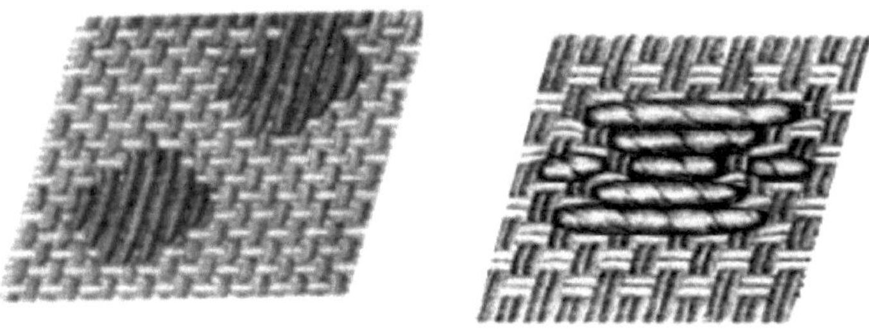

a б

Fig.10. Embroidered fabrics: a - basic embroidery fabrics; b - fine embroidery fabrics.

Conclusion

Leno weaves are produced from yarns and threads of various types of linear density, combinations of type and colour. They are used for making dresses, blouses, curtains, sieves, filters and other products, as well as leno weaves are used for fixing fabric edges on weaving machines. Fabrics of simple leno weaves are produced on weaving machines equipped with a device for setting two warps, as the warp and leno warp tension is different. In addition to conventional looms, these looms have complex looms consisting of two parts: a wing and a half wing. The machines are equipped with an eccentric or carriage yawing mechanism depending on the complexity of the weave. A compensating bar must be installed to compensate for the tension of the leno warp. For the production of complex leno weaves in some cases requires the installation of a specially designed reed and special guides for the movement of the shuttle through the shed. Due to the fact that the dressing and production of openwork fabrics is associated with some difficulties, where possible, they are replaced by fabrics translucent weave, which imitate openwork weaves and are called false ajour. Fabrics of leno weaves are produced from threads and yarns of different types, linear density, combinations of type and colour. They are used for making dresses, blouses, curtains, sieves, filters and other products. The leno weaves are used for fixing fabric edges on hydraulic, rapier and air-jet weaving machines, as well as for fixing edges of narrow fabrics when they are produced in several webs on wide weaving machines.

Embroidery warp fabrics can be produced on any weaving machine of any design with a dobby with a large number of dobbies and the possibility of installing an additional warp for the embroidery warp.

Embroidered weft fabrics can only be produced on multi-colour or multi-weft machines equipped with a special embroidery weft thread device.

LITERATURE

1 Rakhimkhodjaev S.S., Kadyrova D.N. Theory of tissue structure. Textbook. Tashkent. Adabiyot uchkunlari. 2018. - 212 pp.

2 .Kutepov O.S. "Structure and designing of fabrics" - Moscow, Legkaya Industriya 1988.

3 . N. GOKARNESHAN. Fabric structure and design. Copyright © 2004, New Age International (P) Ltd, Publishers Published by New Age International (P) Ltd, Publishers.

4 .P R Lord and M H Mohamed. WEAVING. Conversion of yarn to fabric. Second edition Edited by, North Carolina State University, USA , 408 pages, 1982

5 Damyanov E.B. et al. "Fabric structure and modern methods of fabric design" - Moscow, Legkaya Industriya 1984g

6 E.Sh. Olimboev, "Tukimalar tuzilishi nazariyasi" "Alokdchi" nashr. Toshkent, 2006.

7 Martynova A.A. et al. "Structure and design of fabrics" - Moscow, Legkaya Industriya. 1999г

8 . G. H. Oelsner's. A Handbook of Weaves is the best known and most accessible weaving pattern book. November 18, 2004.

9 .Prabir Kumar Banerjee. Principles of FABRIC FORMATION. © 2015 by Taylor & Francis Group, LLCCRC Press is an imprint of Taylor & Francis Group, an Informa business.

10 .HANDBOOK OF WEAVING *Edited by S Adanur, Department of Textile Engineering, Auburn University, USA* 440 pages 543 figures 68 tables 254 x 176mm hardback 2000 ISBN 1 58716 013 7 Q115.00/US$190.00/Euro160.00

12 .HANDBOOK OF TECHNICAL TEXTILES *Edited by A R Horrocks and S Anand; The Bolton Institute, UK* 576 pages 244x172mm hardback October 2000 ISBN 1 85573 385 4 Q175.00/US$290.00/Euro245.00

13

Printed by Books on Demand GmbH, Norderstedt / Germany